Antique Ice Skates
for the Collector

Russell Herner

4880 Lower Valley Road, Atglen, PA 19310 USA

Published by Schiffer Publishing Ltd.
4880 Lower Valley Road
Atglen, PA 19310
Phone: (610) 593-1777; Fax: (610) 593-2002
E-mail: Schifferbk@aol.com
Please visit our web site catalog at
www.schifferbooks.com

This book may be purchased from the publisher.
Include $3.95 for shipping. Please try your bookstore first.
We are always looking for people to write books on new and related subjects. If you have an idea for a book please contact us at the above address.
You may write for a free catalog.

In Europe, Schiffer books are distributed by
Bushwood Books
6 Marksbury Avenue
Kew Gardens
Surrey TW9 4JF England
Phone: 44 (0) 20-8392-8585; Fax: 44 (0) 20-8392-9876
E-mail: Bushwd@aol.com
Free postage in the UK. Europe: air mail at cost.

Library of Congress Cataloging-in-Publication Data

Herner, Russell, 1939-
Antique ice skates for the collector / Russell Herner.
p. cm.
ISBN 0-7643-1200-6
1. Skating--Equipment and supplies. 2. Skates--Collectors and collecting. I. Title.
GV852 .H47 2001
685'.361'075--dc21
00-010196

Designed by Bonnie M. Hensley
Type set in Bergell LET/Lydian BT

ISBN: 0-7643-1200-6
Printed in China
1 2 3 4

Dedication

This book is dedicated to my wife Marcia, son Mark, and daughters Lori and Jennifer. I am very proud of you for all the things you have accomplished in your careers, and for the respect, love, sensitivity, and help you have given to your family, friends, and everyone around you.

I also dedicate this book to my late mother and father, Art and Gay Herner, who were always there for good counsel, encouragement, guidance, love, and praise. I was blessed with terrific parents.

Acknowledgments

A big thank you goes to my wife Marcia for her patience and understanding over the years in tolerating my "collection" habit. She would regularly say, "We don't need another pair of skates", but after bringing home still another pair, she would eventually come around and forgive me.

To my daughter Lori, I am thankful for your invaluable help in editing the manuscript and your assistance in typing. Your thoughtful suggestions improved the book immensely.

To my son Mark, thank you for your suggestions and help with the computer, and to my daughter Jennifer, thank you for your encouragement and support in developing the book.

I am also very appreciative of my brother Rich, who helped me with some computer software problems during the writing process, and my brother Jim for his help in sorting photos, as well as his instructive suggestions for the text.

My thanks is also extended to Martin Klinkhamer of northern Holland for his translation from Dutch to English of some skating documents, and to Roger Diederen from the Cleveland Museum of Art who translated the Dutch, French, and Latin inscriptions on the 17th century Dutch skating prints.

Lastly I would like to thank the Torrington Historical Society of Torrington, Connecticut, for the data provided to me on the Union Hardware Co.

Contents

Foreword

I was recently given the privilege of reviewing Russell Herner's manuscript for this book. I've known Russ as a friend and fellow antique tool collector for many years but had no suspicion that he also collected antique ice skates and skating lanterns! Obviously, he is very knowledgeable about the subject and has done a monumental amount of research during his travels, both in this country and in Europe, while in search of these artifacts.

During this review, I was struck by the diversity, ingenuity, and evolution of skate design over the centuries. Even the earliest bone skates display the individual artistic talent of the maker. As antique tools have fascinated me, so have the products of those tools. Coincidental to this book, at an auction just two months ago, I purchased a "box lot" of three pairs of wooden bodied skates (my first). One pair was a child's size and measured just 7 1/4 inches overall! All were "manufactured" in the 1800s, and I marvel at their beautiful styling, design, and fine workmanship. Am I hooked on skates? Possibly!

This book is believed to be the first publication ever dedicated exclusively to ice skates and skating lanterns. It contains a wealth of information for the collector, as well as for those with even a causal interest in skating and skating paraphernalia. The beautifully executed photographs of the skates, the skating lanterns, and the related artwork will help provide many, many hours of pleasure for the reader. In addition, the informative price guides will greatly benefit the beginning and the avid skate collector, as well as the dealer.

Ronald W. Pearson, D.O.
Antique Tool Collector

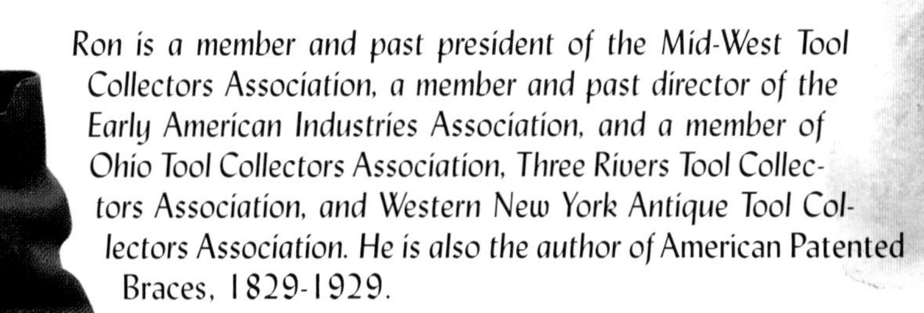

Ron is a member and past president of the Mid-West Tool Collectors Association, a member and past director of the Early American Industries Association, and a member of Ohio Tool Collectors Association, Three Rivers Tool Collectors Association, and Western New York Antique Tool Collectors Association. He is also the author of American Patented Braces, 1829-1929.

Introduction

Antique ice skates have been highly sought after for many years by collectors from all corners of the globe, but to my knowledge there have never been any books written on ice skate collecting. There is a definite need for a publication to assist the collector on identifying the various skate types, styles, designs, approximate ages, countries of origin, and rarity. A general price guide range would also be helpful.

I have struggled with these questions over the years and have attempted in the text to address them. Since this ice skate collector's book is a first, it is only my interpretation of the skate's history, description, and origin. My conclusions were drawn after thirty-five years of ice skate collecting. It is my hope that this will be a good start for all the antique ice skate collectors. They can now come forward and add their knowledge to the work that is presented in this text. With the combined knowledge of the current collectors and further study of additional skate examples, a better understanding of the chronological history of skates and their manufacturing can be developed.

There are many antique ice skate collectors around the world, but I feel most of the collectors have never communicated with one another and should. Perhaps this book will stimulate someone to come forward and organize an Antique Ice Skating Club. It would unite the collectors in a common way by preserving ice skates and enabling members to share with each other their knowledge and collections. Members could also freely swap, sell, and trade duplicate skates. And most importantly a club would allow future generations to be exposed to an almost forgotten American heritage. What would it be like to attend an old fashioned ice skating party on someone's pond complete with bonfire, authentic dress, antique ice skates, and steaming cups of hot chocolate? A club could include all these possibilities and provide a unique opportunity for the collectors as well as their families.

Chapter 1

A Brief Overview Of Ice Skates

Their History, Origin, Types, Craftsmanship, and Construction

The origin of the word "skate" probably came from the German word "Schake" or "Shank" meaning a leg bone of an animal. The earliest ice skates were made from large animal bones with holes drilled through each end of them to accommodate leather thongs. The leather thongs were then used to attach the bone skates to the shoe or foot. The animal bones selected to fashion the skate runners were taken from the legs and shinbones of elk, cows, sheep, and horses generally, but other animal bones were used as well. See *Figure 1*.

Figure 1. This ice skate made from an animal bone is the first known and is dated circa 1300. Leather thongs were passed through holes drilled on either end of the bone to fasten the skates to the boots. The bottom of the bone was scraped smooth for the runner to pass over the ice.

The first use of ice skates as a means of transportation is very difficult to pinpoint, but references date bone skates as far back as 1500 years. Bone skates have been found in several northern countries and are difficult to date, though they are probably a minimum of 700 years old. The metal skate blade, which followed the bone skate, was known to have existed in Holland around 1300. To my knowledge no bone skates have ever been radio carbon dated though that would date the animal bones quite accurately.

There is an early twelfth century reference in England that bones were tied to people's feet and served as ice skates. It is from an early document by Fitzstephen titled "Description of London." It was written in Latin and published in 1180. It is found in a book titled *Skating*, by J.M. Heathcote, published in 1892

> When the great fenne or moore (which watereth the walls of the citie on the North side) is frozen many young men play on the yce…some striding as wide as they may doe slide swiftle; as some tye bones to their feete and under their heeles, and shoving themselves with a little picked staffe do slide as swiftlie as a birde flyeth in the aire or an arrow out of a cross-bow.

There are several references to early skating in Holland, Europe, and the Scandinavian countries, but it is quite difficult to determine exactly where ice skating originated. Some historians feel ice skating started in one of the Scandinavian countries such as Norway or Holland, or even in several other countries simultaneously. No one will probably ever know for sure unless some archaeological discovery is made that will prove its origin, but even then it will be debated. Though specific conclusions cannot be drawn with regard to its advent and location, a general chronological order of ice skate development around the world can be described.

Obviously skate design and development varied in different parts of the world at the same time due to community development, local customs, and the natural resistance to change to a new skate design. Therefore there are no clear dates established for the transition from one design to the next. Wooden platform skates were sold and used in one country at the same time metal skates were being used in other countries, with design changes being made over many years in various regions of the world. That being said, there is a general progression of skate development:

- The first skate runners were made from bones taken from a variety of animals and attached to the foot by leather thongs.
- Wide wrought iron blades were made and attached to the boot as skates.
- Thinner wrought iron blades were developed as runners for the skates and mounted vertically. They were attached to the upper wooden platforms in a variety of ways. The platforms had two or three slots morticed in them allowing leather thongs to pass through for fastening to the upper shoe or boot.
- An all metal ice skate was developed, both runner and platform. These all metal skates (also known as "club" skates) were attached to the shoe by adjustable metal screw clamps. E.V. Bushnell invented the first integral all metal foot plate and blade skate in 1848. A few of the all metal skates were still fastened by leather straps as the earlier wooden models were.
- An all metal skate runner and a platform with holes drilled in it to screw or fasten the skate to the bottom of the shoe. The metal skates were bought and then screwed to boots or shoes.
- The last development in ice skate design incorporated metal bladed runners fastened permanently to the upper shoe with rivets as some of the less expensive skates are today.

Over time there was a variety of experimentation with skate fastenings to the boot. Leather straps or thongs were originally used to fasten the wooden platforms to the shoes, but the skates would come off or they would get loose and have to be frequently retightened again; a real pain! Wood screws were then mounted upside down in the heel and sharp spikes were mounted on the toe area to help keep the shoe from sliding or coming off the skate platform. The development of the various stages of ice skates will be examined in further detail and many of the skate designs and means of fastening the skate to the foot will be illustrated.

Two Basic Types of Skates

We will be addressing two basic types of skates, early figure skates and early speed skates, along with a variety of specialized skates. Both categories include early handmade skates and factory made skates.

The figure skates, *Figure 2*, are shorter and have a curved profiled runner on the bottom that allows approximately 2" inches of the runner to touch the ice

Figure 2. An unusual pair of handmade figure skates with a curved profile runner on the bottom has platforms made of tiger stripe maple, the same wood as in a fine violin. Note the large screws in the heels and sharp spikes in the toe area to keep the boot from slipping off the platform. Leather straps and brass buckles were used for fastening. Length 11 5/8", Width 2 3/8", Platform height 1 3/4", c. 1850. Class F: $500-750.

surface at any one time. The speed skates are generally longer with a straight or flat-bottomed runner, such as a Dutch racing skate, *Figure* 3.

Because early skates were the subject of much experimentation, the handmade skates included a variety of unusual designs. Many different materials were used that were available at the time. *Figure* 4 shows a good example of a handmade pair of skates.

Figure 3. Dutch speed skates featuring long runners and a nice turn-up curl on the front. Note the decorative touch added to the top of the runner near the platform. These skates were probably made in Germany and imported to Holland for use. Leather thongs were used in the rear holes to tie around the ankles and straps were used to hold the toe on the front. Triangular spiked nuts were used on the heel and toe to keep them from slipping off the platform. Length 17", Width 2 1/4", Platform height 1 1/2", c. 1870. Class D: $200-350.

Figure 4. An excellent example of a pair of handmade skates. The blacksmith hand forged the runner blades out of an old "file"! The front turn up is made from the tang of the file and the rear attachment eyelet was forged from its blade. Note the distinctive file teeth on the side of the skate runners. The wooden platform was attached with a pin on the front. The old leathers are still attached for the fastenings. Length 15", Width 2 5/8", Platform height 2 1/4", c. 1850. Class E: $350-500.

Opposite page

Figure 5. An exceptional pair of English figure skates with rosewood platforms. There are two brass inlaid decorative foot plates on the heel and toe. Screws in the heel and spikes in the toe area prevented the boot from slipping off the platform. Three leather straps were used to fasten the boot. They were manufactured by the James Howarth Co., Sheffield, England. Length 14", Width 2 3/8", Platform height 1 1/2", c. 1860. Class E: $350-500.

The factory made skates were also manufactured in several designs and styles. Many of the companies took pride in making their ice skates with the highest quality possible. Factory made skates often were of outstanding design with brass inlays and exotic hard woods. See *Figure* 5.

Craftsmanship in Skate Making

My interest in antique ice skates came through the association of collecting primitive craft tools and antiques in general. After collecting old tools for several years one acquires an appreciation of the skill and craftsmanship required to make the quality heirlooms we collect today. Each craftsman, such as the carpenter, cabinetmaker, blacksmith, cooper, harness maker, leather worker, woodcarver, wheelwright, wagon maker, clockmaker, pewterer, silversmith, etc. possessed a considerable amount of skill and specialization in his craft. This skilled craftsmanship was developed over a period of several years in the apprenticeship programs.

A great number of today's hand made, highly collectable antiques were made by a combination of these skilled craftsman. Early ice skates would be a good example, requiring three crafts: a blacksmith for the hand wrought runners, a woodcarver for the platforms, and a leatherworker for the leather fastenings. A seventeenth century skate maker in Holland, for example, would probably employ a few craftsmen in his shop who specialized in these three different crafts. They would put together a high quality pair of skates using the special skills of each worker.

Occasionally one man would be adept in all these crafts and would make the skates himself. But even if one were adept in all these crafts today and attempted to make a quality hand wrought pair of skates, it quickly would become apparent that it is a big challenge. After completing a pair of skates with blacksmith wrought curled runners, carved platforms with morticed slots, and hand stitched leather straps, the pair of skates in the antique shop would no longer seem overpriced, figuring all the hours involved! They would be a bargain.

Several years ago I ran across a nice pair of handmade ice skates and I decided, then and there, that I had to own them because of their beauty and craftsmanship. This was the beginning of my collection (or the beginning of a disease that I caught). Over a period of years I collected skates that have different designs, character, styles, and fastenings. Most collectors will always find that different pair of skates that they must own!

Many of the early ice skates are examples of outstanding art. The blacksmith's talent is revealed in the graceful curved front of a hand wrought runner ending with a tight reverse curl. His smooth hammer marks may still be seen on the runner blades. The woodcarver's skill is demonstrated in the wooden platforms carved in a variety of shapes to fit the foot. The leather worker would skillfully hand stitch the leather straps and buckles together. A pair of large curled skates is very appealing to the eye and definitely a work of art, as seen in *Figure 6*.

Figure 7. On top of the blacksmith's anvil is the skate runner being held with a tong and waiting to be forged into a curl shape. The blacksmith would heat the straight runner to a cherry red color and then forge the curl diameter over the horn of the anvil with the hammer. The art and skill of the blacksmith was mastered over many years of working with the forge, anvil, and hammer. He had to know the metal's characteristics and the proper temperatures for forging and bending the iron into shapes. The highest requirement of the blacksmith was the knowledge and skill to forge weld two unlike pieces of metal together merely by hammering them on the anvil. The several examples of the ice skate runners forged to the stanchions demonstrate this skill.

Figure 8. An old pair of skates with the upper leathers sewn and fastened to the platforms. Skate parts are shown in the background along with a few leather tools used by the harness maker/leather worker. Note the stacks of leather straps, on the right, in various widths to be later crafted into toe straps for the skates.

Opposite page

Figure 6. An outstanding, one-of-a-kind pair of handmade ice skates. They are a "work of art." The runners were gracefully hand forged from rectangular stock into a small diameter. The exceptionally large curl continues to curl into a diamond-shaped finial at the tip. The curl diameter is the largest in my collection, measuring 6 5/16" from the ice to the top of the curl. The platforms are made of walnut. Length 13 3/4", Width 2 1/4", Platform height 2", Curl diameter 6 5/16", c. 1860. Class H: $1000-1500.

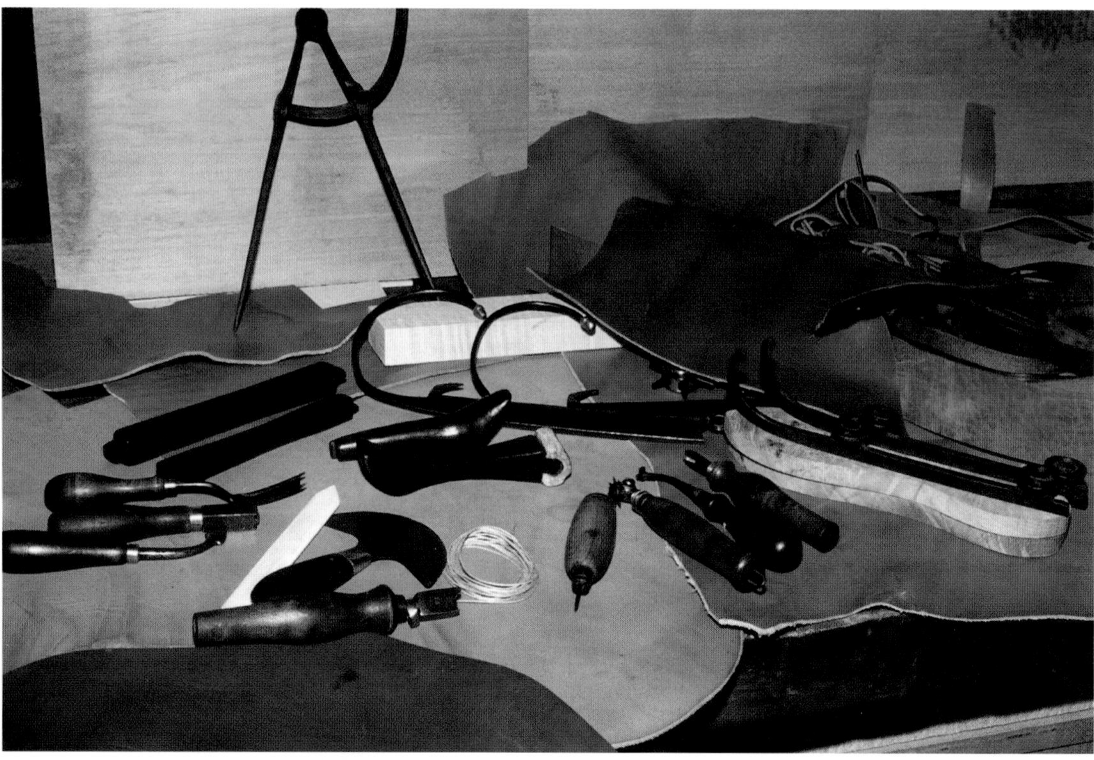

Figure 9. Several tools, such as awls, creasers, burnishing tools, and unassembled parts of the skates, which were used by the harness maker or leather worker in crafting ice skate fastenings. Note the burnishing tools, in the center of the photo, used to smooth the seam joints down. The burnishing tools were made of very hard lignum vitae wood. One of them has a bone wear piece on the end of it. The tanned leather hides shown are ready to be crafted into boot uppers, harness straps, and a variety of other fastenings for the early skates.

Figure 10. The leather worker requires a variety of tools such as these. The skate runners shown are ready to be assembled and leathers attached. The handled tool or stitching wheel in the foreground with the star wheel on it was used to put a stitching pattern on the leather. The stitching wheel was pressed and rolled along the leather leaving uniform prick points as a guide to stitch evenly with the needle and thread later. The awl was used, as shown in the center right, to punch holes through the leather. The T-handled tool in the foreground was a leather cutter with a sharp edge on the radius. A pair of skates with leather uppers is shown along with straps and linen thread on the right.

Figure 11. A few of the tools used by the woodworker and carver to fashion the wooden platforms for the skates. A German goose wing broad axe, used to hew or square up the logs, is shown in the background. A French plow plane, to the upper right, was used to cut a groove in the platform into which the runner was set. A smoothing plane in the shape of a dove, along with a horn plane, are shown in the center left. Wood stock in the vice is ready to be sawed into the foot shape for the platform.

Figure 12. Two bird's eye maple platforms sawed out in the shape of the skate platform. They will be drilled to inlet the round stanchion mounting washers. Large curled runners with the clamp that goes up into the front of the platform are shown. A small cherry horn plane with decorative carving and dated 1773 is shown to the left.

Figure 13. A variety of carving chisels was used to carve decorative designs on the skate's platforms, notches for stanchions, or brass inlay plates. The chisels have rosewood handles and were made by the S.J. Addis Co., London, England. The company mark, stamped on the tang, is the Masonic emblem, a square, and compasses. A spoke shave is also lying on the bench to round and smooth the edges of the platform.

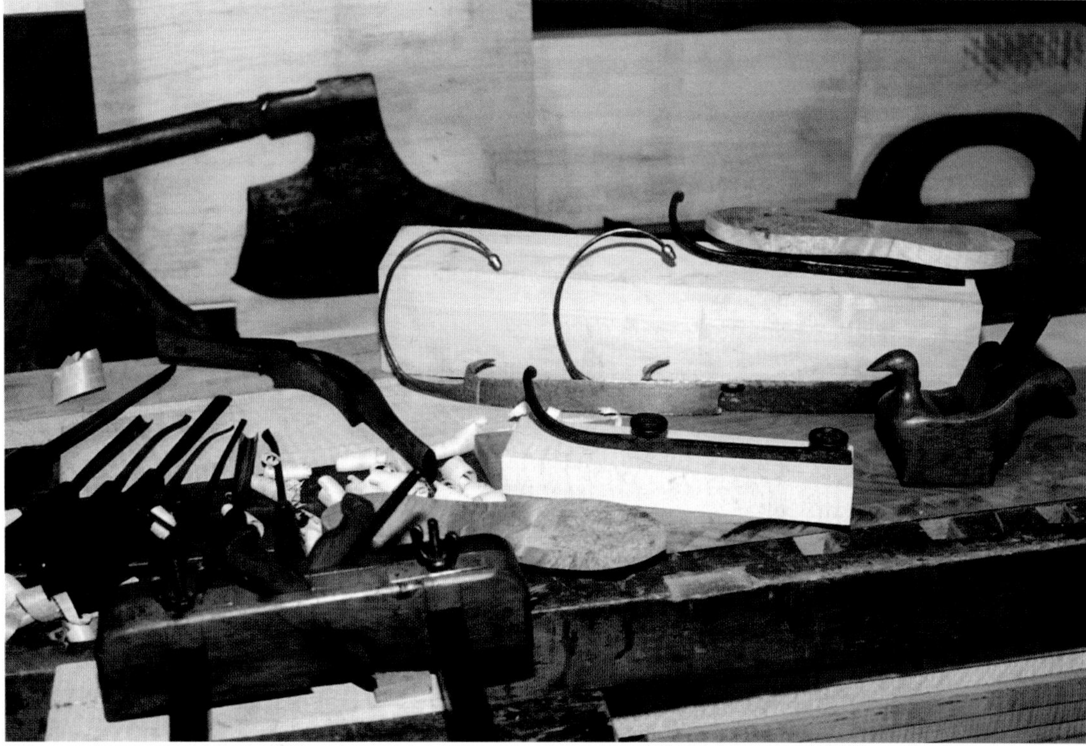

Figure 14. A tongue and groove plane was used to cut a groove in the wood platform into which the skate runner would fit. The small iron runners are shown loose before they have been fitted into the platform. A wooden brace in the background and chisels to the left are shown.

Figure 15. Some of the tools in a typical woodworker's shop that would have been used in the making of skates during the 19th century. On the back wall is a wooden T-bevel square, calipers, a mortice gauge, square, compasses, a small square, and double calipers. On the bench in the background is a bench plane, horn plane, and chisels. Center left is a rosewood plow plane with ivory tips along with some skate parts.

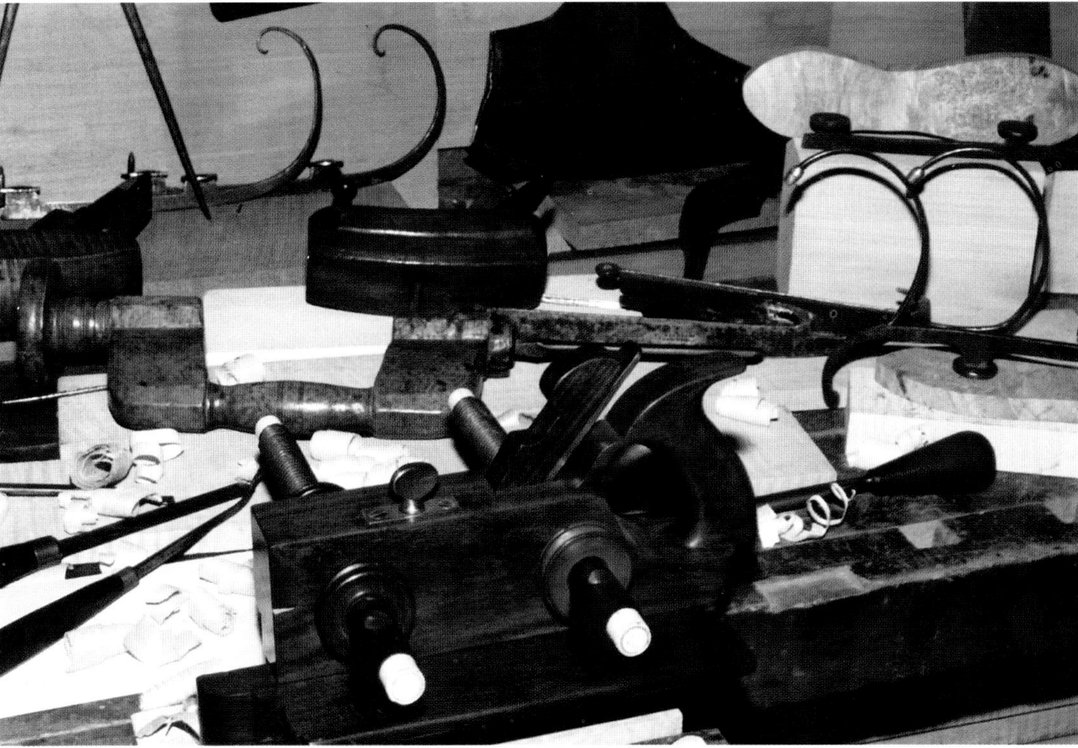

Figure 16. A first class plow plane in solid rosewood embellished with four ivory tips, is in the foreground. The plow plane would have been used to cut the groove in the platform for the runner blade. Behind the plow plane is a solid burl wooden brace for drilling holes. A pair of hand forged curled runner blades are on the bench ready to be fitted. A tiger stripe plane, horn plane, chisels, and skate parts complement this photo. The tools shown would be on the high end of quality in skate making.

The Parts of a Handmade Skate

A handmade pair of Figure Skates is comprised of several parts or features. Refer to the drawing in *Figure 17* for each part described.

Wooden Platform. The top wooden part of the ice skate is called the "platform." Platforms were made in a variety of woods, although most were maple and beech. Some of the higher quality skates were made from walnut, tiger maple, bird's eye maple, cherry, and rosewood. Generally the wooden platforms were made from flat stock, but others were carved in an arch pattern conforming to the arch of the foot. In a plan or top profile view the platforms were sawed out in the shape of the foot.

Morticed Slots. Morticed slots were carved across and through the wooden platform to accommodate the leather straps. The straps were used to fasten the skates tightly to the shoe with buckles. There were generally two or three morticed slots in the platform depending upon the design of the skates.

When two slots were used, they were located at the heel and toe area of the platform. One strap was then used to tightly fasten the toe, and the other strap was wrapped up around the ankle and secured with a buckle. When three strap slots were used, the front and center slots allowed a crisscross pattern of the straps, which secured the toe area. The back slot was again used to strap and secure the ankle. There were always exceptions to these rules of fastening the skates to the foot.

Leather Straps. Leather straps were used to fasten the skates to the skater's shoes. They were threaded through the morticed slots in the platform and tightened or fastened down by metal or brass buckles.

Support Nails on Top of Platform. Many times sharp nails or screws were located on the top front of the platform to keep the toe area of the shoe from sliding sideways or slip-

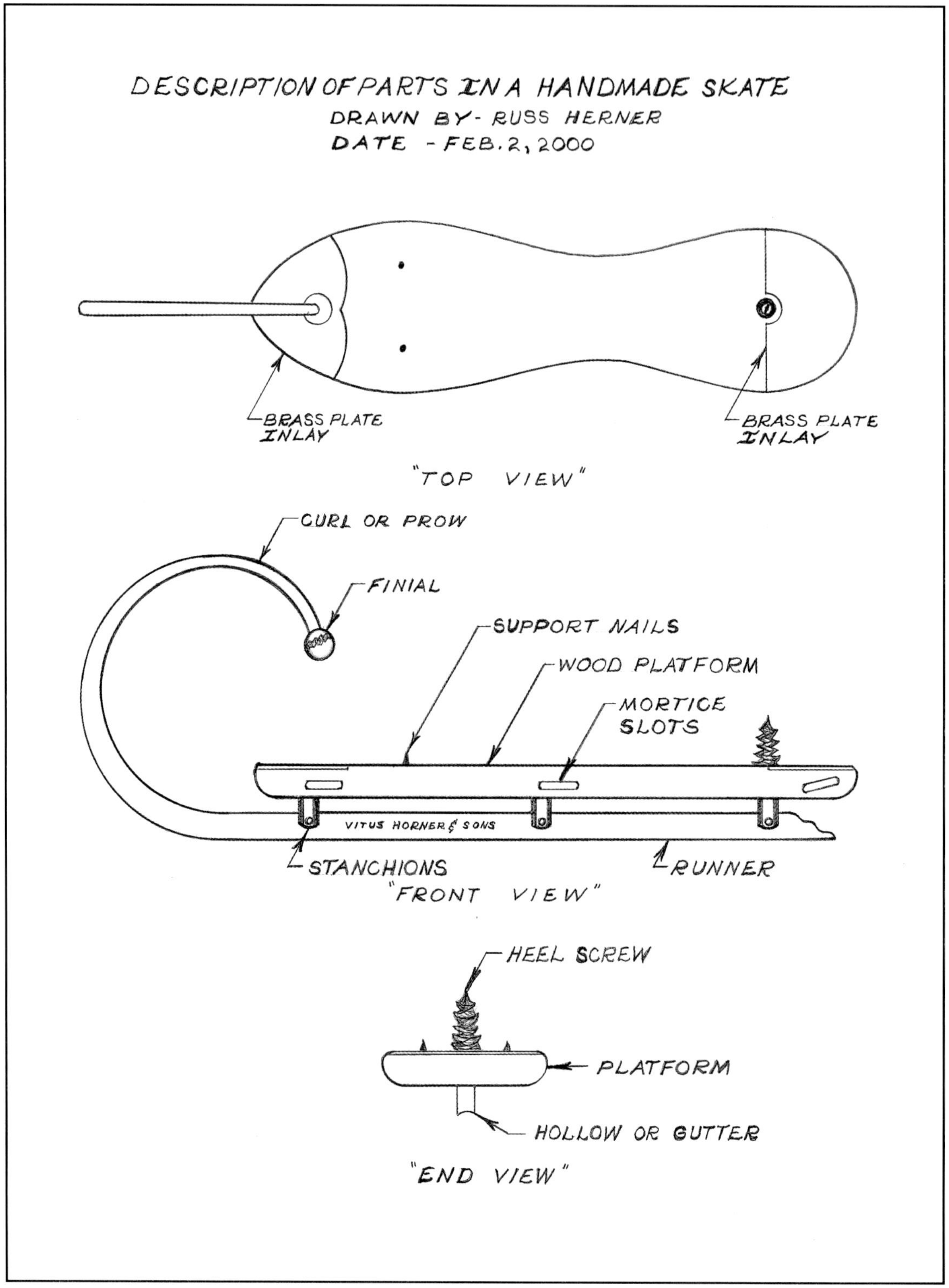

Figure 17. For the benefit of the reader, this is a drawing illustrating all of the descriptive parts of a homemade pair of ice skates.

ping off the platform. These sharp nails protruded above the platform surface about 1/8" inch. They were made in a variety of shapes but all were quite sharp to dig into the bottom of the shoe to secure it

Heel Screw or Spike. Most of the early wooden platform skates had a screw sticking up through the center of the heel, which was screwed or turned into the heel of the skater's shoe. This screw kept the heel from sliding off the platform.

The screw length varied from about 1/2" to 1-1/4" long. The lower portion of the screw was sometimes threaded so it could be adjusted up or down. After the screw was adjusted to length, it was locked in place by a retainer nut from the underside of the platform.

Some skates had a sharp spike or triangular shaped heel spike that was inserted into the heel and served the same purpose as the screw. Many of the factory-made skates had threaded replacement metal spikes of varying lengths for the heels. Most of the spikes were generally shorter than the screws used. A person had better know how thick his heel is before screwing a 1" inch long wood screw up into it. If not, he might think he had a bone spur on his heel!

Metal Runners or Blades. The metal runner or blade was originally hand wrought into shape by the blacksmith on the anvil. He forged the blades into a variety of thicknesses ranging from about 1/8" to 7/16". The average thickness of a runner was between 3/16" and 1/4".

The blacksmith sometimes forge welded a narrow piece of harder steel to the bottom of the wrought iron skate runner. The harder steel piece functioned better because it would keep its sharpness much longer. The same principle was used in forge welding a band of harder steel to the bottom of the early goose wing broad axe blades. They would stay razor sharp much longer during the process of hewing square wooden beams from the round logs.

The early skates were forged straight or flat on the bottom. Later the figure skate was developed with the runner having a slight circular profile on the bottom. This curve allowed approximately 2" inches of the runner to contact the ice surface at any one time.

A few of the handmade runners were formed out of old worn out files or rasps. Metal was scarce in the early development of our country and nothing was thrown away, especially an old file that could be forged into another useful product. The old files were very hard iron and served quite well as good skate runners that kept their sharpness longer.

A few years ago I asked an old gentleman about buying some old curled runner skates and he referred to them as "butcher blades curled skates." It was the only name he knew them by. I suppose that a few old, thick-bladed butcher knives that were worn down from sharpening over the years could have been reforged into a nice pair of skate runners. Again, nothing was thrown away.

Many of the later factory made skate runners were made from cast steel. A great number of the advertisements stated that they were using the "best" cast steel available for their runners.

Hollow Ground Runners. Many of the blade runners had a U-shaped groove ground into the bottom, which was called a "hollow ground" runner. This feature is also known as a "gutter" and was developed around 1870. Patent No. 37169 dated Dec. 16, 1862 incorporated a groove on each edge of the under or supporting surface of the runner giving it a proper hold upon the ice. Is this the original invention or idea of the gutter? See *Figure 18*. This gutter shape on the

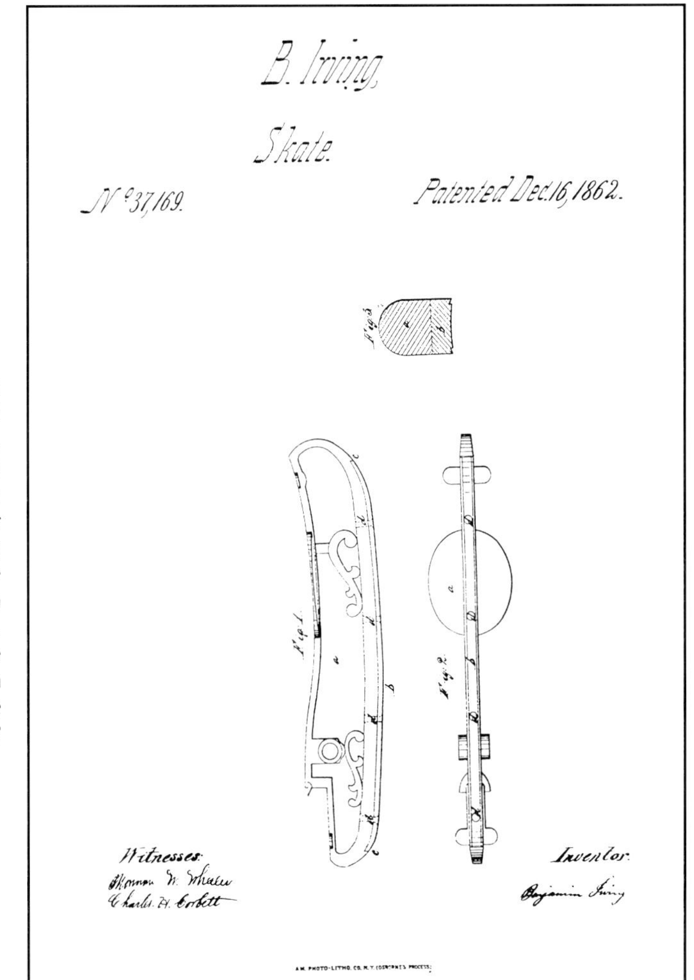

Figure 18. Patent No. 37169, dated Dec. 16, 1862, incorporated a groove on each edge of the under or supporting surface of the runner, giving it a grip on the ice. Could this be the original invention or idea of the hollow or gutter?

bottom allowed a better bite into the ice giving the skater more stability and security while skating. A few runners were made by bending over a thick piece of sheet metal into a "U" shape, forming a double runner on the ice, thus simulating the gutter feature.

Curl or Prow on the Runner. The curl or prow on the front of the skate runner was designed and used by the Dutch to allow the skates to move more easily over the rough ice. The Dutch curls were made in a variety of sizes or diameters. The larger the curls, the more desirable they are to collectors today. The largest curl example in the writer's collection is 6 3/8 inch in diameter, which is exceptionally large. It is measured from the bottom of the blade to the top of the curl.

Decorative Finial at the Tip of the Curl. On many large curled skates the maker added a decorative finial or forged a reverse tight curl. The decorative finial added a great deal of class to the overall appearance of the skates.

A few of the manufacturers of skates added brass acorn finials at the tip, and others were in the shape of a bell. The blacksmith would forge a variety of shaped metal finials on the end of the curled tip as well. Some of them were forged into a graceful, scrolled, curled tip of a couple revolutions. This enhanced the looks of the skates appreciably.

Stanchions. The stanchions of the skate are the vertical support pillars connecting the upper platform to the lower metal runner. Generally there are two or three stanchions on a skate. The stanchions were forged as part of the bottom runners in some cases and in others they were separate pieces forge welded to the runners. Some were riveted onto the skates while others were screwed. Many of the manufactured skates used turned brass pillar stanchions. They came in a variety of shapes and materials. A few elegant ones were made of brass in the shape of bells. The tops of many stanchions were threaded with a round nut securing them flush with the top of the platform.

A few skates did not have any stanchions on them at all. The platform was attached to the front and rear of the runner only, similar to a spring supported platform.

Chapter Two

Holland

Motherland of Ice Skating

For most people the Netherlands or Holland is associated with tulips and early ice skating. About 18% of Holland's territorial area is water, comprised of large arms of the sea, lakes, rivers, canals, ponds, and ditches. A good freeze in Holland ensures plenty of ice on which to skate.

In earlier days in Holland, wet cold winters caused the mud roads to become very bad and almost impossible for travel. Ice skating was almost a necessity for getting around during the frozen winter months. The Dutch people used the frozen rivers and canals to skate to their friends' and neighbors' houses, as well as to the surrounding villages. Ice skating in Holland also served as a means of communicating the news between the villages. It is said that many skaters traveled great distances daily on the ice, several miles in many cases.

It is generally accepted that the Dutch people were the inventors of the first iron runners on skates. They were using iron bladed skates in Holland as early as the 1300s.

An early woodcut printing by Johannes Brugman in 1498, at Schiedam, Holland, illustrates the fact that women participated in the ice skating sport early on, and that metal skates were in use at that time. It captured the event of 1396 when Lydwina, a young teenager, broke her rib when she was hit by another skater and knocked down on the ice. She suffered for years with complications from the accident and never recovered. After her death she was adopted as St. Lydwina of Schiedam, Holland, patron saint of skaters. The printing also shows a male approaching the accident on a pair of skates, pushing sideways with his feet as a modern skater does with metal skates.

Ice skating became one of the earliest sports in Holland. By the middle of the seventeenth century, skating in Holland was a national pastime and a competitive sport, see *Figures* 19,20,21,22. In Holland the entire family skated, as seen by the two small children in the Dutch painting in *Figure* 23. They are sitting on a log

Figure 19. An early Dutch scenic print depicts skaters, a decorated horse and sleigh, pic sleds, hockey players, and an ice boat. Note one skater putting his skates on and another one looking up to the sky after a fall. Note the golf game being played on the ice. A thatched roof building and windmill are shown in the background. The translation of the Dutch inscription below the print is: "Let's have fun now, riding with the ice sled Let's ice skate, we want to play golf, now that it's such healthy air." The title appears to be "January Air." Pieter Nolpe designed and engraved the image, 1650.

dressed up in typical Dutch costumes including a hat and white bonnet. They seem to anticipate getting up to skate on the frozen canal. There are two adults crossing a foot bridge with a windmill and orange tiled roofs in the background. The painting was not signed.

Figure 20. Another early Dutch scene illustrates several skaters, some on their backs, and a man on a pic sled with poles pushing himself along on the ice. On the left is a Paard En Arreslee, or horse and sled, all decorated and capturing everyone's attention. In the center is a lady with a wooden barrel mounted on a push sled selling what is probably cider. They all seem to be dressed in their finest clothes and enjoying the afternoon activities. 18th century.

Figure 22. This early Dutch scene shows a skater pushing a sled with wooden barrels on it. Another skater is retightening the leather fastenings on his skates at the bench. A third is pulling his wife along on a sled, as it should be! Note the church, windmill, and houses in the background of this small Dutch village. 18th century.

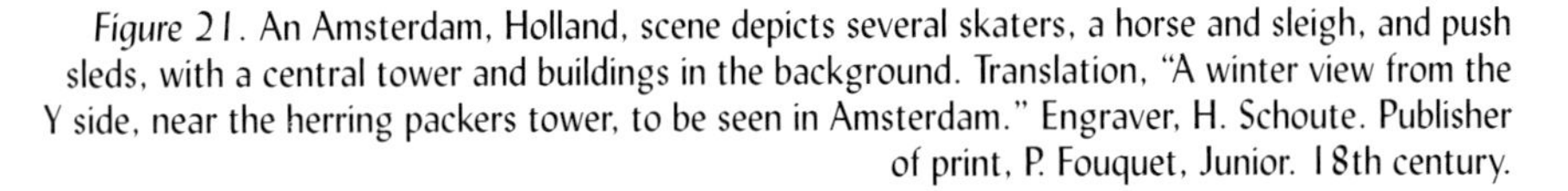

Figure 21. An Amsterdam, Holland, scene depicts several skaters, a horse and sleigh, and push sleds, with a central tower and buildings in the background. Translation, "A winter view from the Y side, near the herring packers tower, to be seen in Amsterdam." Engraver, H. Schoute. Publisher of print, P. Fouquet, Junior. 18th century.

Figure 23. Everyone gets into the ice skating sport in Holland, even children, as depicted in this Dutch painting on canvas. Note the nice large curls on the children's skates. Painting measures 25" x 48". c. 1880.

The Dutch skate makers developed the distinctive large curl or prow on the front of the runners to help the skaters glide over the rough ice on the rivers, canals, and ditches. These large curled runner skates are very desirable and are sought after today by the collectors around the world. Germany made these large curled ice skates as well, and exported them to Holland and other countries. It has been said that the large curls on the Dutch skates in Holland were prohibited around 1860 because they became too dangerous to the skaters during a fall. The curl diameter was then reduced in size and attached even with the top of the platform, thus making a safer skate. Some of the small curled runners were enclosed or sandwiched inside the wooden platform. The Dutch speed skates were generally very long, with flat runners so they could more easily speed skate straight ahead and travel long distances.

As a note of interest, in the last few years thousands of wooden Dutch skates have been flooding the U.S. market that date from the 1960s and 1970s! Many have been sold as antiques! One must be careful in buying ice skates, to ensure that they are truly antique.

Speed skating soon became a competitive sport in Holland and official races were established throughout the country. In 1840 the town of Dokkum in the north Friesland Province of Holland was the first town to set up a skating association. Several races have been set up around the other Provinces, but the best known one is " Elfstedentocht," or the Eleven Towns Race, held in Friesland. Its roots date to 1890 when a skater raced through each of the eleven towns in a time of 12 hours and 55 minutes, and then recorded his signature as proof he was there. The Eleven Towns Race was founded in 1908 and had its first official race in 1912. The race starts about 5:00 A.M. in Leeuwarden and goes through Leeuwarden, Sneek, Ijlst, Sloten, Stavoren, Hindeloopen, Workum, Bolsward, Harlingen, Franeker, and Dokkum in a circle covering 125 miles!

When the race is held during blizzard conditions it is very challenging and most racers never finish. Photos of the racers' faces show them completely covered with ice. The races bring out about 16,000 of the world's best competitive long distant skaters and over 500,000 spectators!

I had the privilege to meet a nice gentleman, named William Augustin, in the Hindeloopen skate museum. He was very knowledgeable, and served as my guide and translator at the museum. He participated in the Eleven Town Races held in 1941, 1942, 1947, 1954, 1956, 1963, 1985, 1986, and 1997. He was 75 years young during the 1997 race! He conducted me on a very enjoyable tour around the quaint little town of Hindeloopen where we stopped at the fish market and sampled raw herring!

Figure 24. Dutch engraving by Jan Luikjen titled "The Skate Maker, 1694." This scene illustrates an early Dutch skate maker shaping the wooden platform of the ice skate. Note the other worker in the background. *Courtesy of the Rare Book and Manuscript Library, Columbia University, New York, New York.*

The latest speed skate design was developed in Holland and is called the "Klapschaats." It is a skate that has a spring-loaded hinged front runner that allows the skater to just partially lift the weight of the entire skate on each stride. It apparently saves the skater a little energy on each stride and thus pinches a bit more speed and endurance out of the skater and gives that edge on long distance races.

At the 1999 World Speed Skating Championship races at Heerenveen, Holland, which I attended, the racers were all using the Klapskates, which one could hear clicking and clapping on every stride as they went around the rink. The speeds that the racers attain is unbelievable when witnessed live and a few feet off the ice. I had the privilege to see Holland's Marrianne Timmer race, and she won the Worlds 1500 meter race the following day! It was quite an experience and treat for me to see the best racers in the world compete.

Among the old skate makers in the Friesland area of Holland are J. Nooitgedagt, A.K. Hoekstra, G.A. Ruiter, and Hamstra's.

Figure 25. A pair of nice long Dutch skates with beautiful brass finials at the tips of the runners. The platforms are made of walnut with a spiked nut on the heel to secure the boot. The Dutch maker's name is stamped on the runner, Romijn from Delft. There are two leather straps in the morticed slots for the fastenings. Note the Dutch wooden shoes called klompen and the tulips in the background. Length 17", Width 2 1/4", Platform height 1 1/2", Curl height 3 3/4", c. 1865. Class D: $200-350.

Figure 26. Early Dutch skates with nice heart-shaped decorative finials at the tip of the curls. The runners are hand wrought and terminate in the center of the heel. The platform shape is rectangular with straight sides continuing to the heel. They are Frisian skates from the northern province of Holland called Friesland. The heel nut has spiked teeth to hold the boot on the platform. There are two morticed slots for the leather fastenings to go through. Length 16 1/4", Width 1 7/8", Platform height 1 1/2", Curl height 5 1/4", c. 1830. Class D: $200-350.

Figure 27. These Dutch skates have decorative finials at the tips of the runners. The runner stops in the center of the heel at the rear. The old leather straps are still attached through two morticed slots in the platform. The heel nut has spikes on it to keep the boot from slipping off the platform. The runner curl has a safety feature incorporated in it, with the metal curl sandwiched between the wooden platform to help protect the skater in a fall. Length 12 1/8", Width 1 3/4", Platform height 1", Curl height 2 3/8", c. 1890. Class C: $100-200.

Figure 28. Dutch skates with wooden shoes mounted on them. The skates have beautifully turned brass finials at the tips. The platforms are made of walnut wood with a small metal cup or clip at the rear to keep the boot from sliding off. The curl is again sandwiched between the wooden platform. The heel cup leather straps are still intact. Note the old wooden shoes have worm holes in them, and someone, possibly a shoemaker, later fashioned a pair of leather tops to them with shoemaker' s wooden pegs serving as nails. It is quite the combination! Length 15 1/4", Width 2 1/4", Platform height 1 1/2", Curl height 3 3/8", c. 1870. Class D: $200-350.

Figure 29. A pair of Dutch skates with large brass cast heels mounted on the platform with copper nails. The platforms are made of walnut mounted to the runners with copper rivets and brass diamond-shaped washers. The runners are in the shape of an hourglass and are very narrow in the center. The toe leathers are very wide, covering the foot area and tightened by laced strings. The leathers are fastened to the platform with brass-headed nails. Length 12 1/4", Width 2 1/2", Platform height 1 5/8", c. 1890. Class D: $200-350.

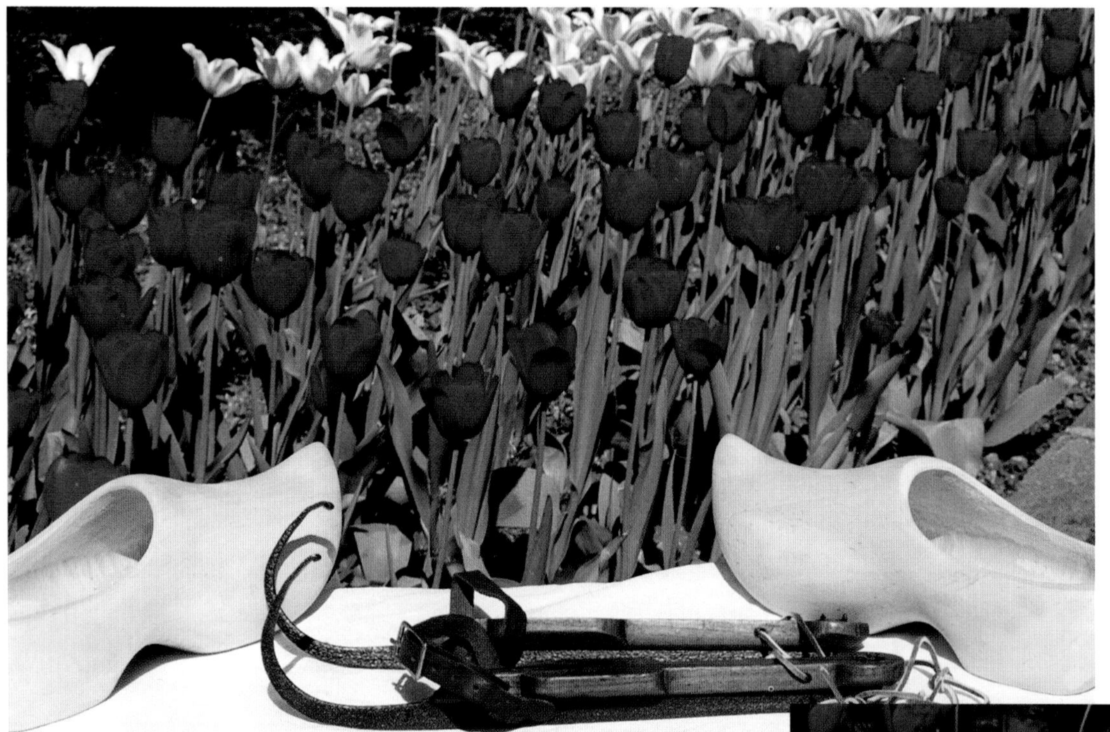

Figure 30. These very early Dutch skates have hand forged runners and nice curls with finials. The platform is sawed out wider at the toe and heel and narrower at the arch. The front fastenings have leather straps and at the rear are leather thongs to tie around the ankle. The heel screw has three barbs on it to prevent the boot from slipping off. These skates are a bit rare. Length 15 3/8", Width 1 3/4", Platform height 1 1/4", Curl diameter 4 1/8", c. 1800. Class E: $350-500.

Figure 31. These Dutch skates have hand wrought runners and nice turn up curls. The top of the runner has a decorative design near the platform. The platform has old ropes on it for fastening, instead of the original leather straps. The heel screw has three barbs to prevent boot slippage. It is a possibility that these skates were made and imported from Germany to Holland. Length 13 5/8", Width 2", Platform height 1 3/8", Curl diameter 3", c. 1870. Class C: $100-200.

Figure 32. Handmade skates with nicely forged curls on the front with bent over finials. The skate curl certainly has a Dutch look but the skates may be American made. The runner has two stanchions instead of a solid blade like a Dutch skate. The heel screws were not often used on Dutch skates either. Possibly they were made by a Dutchman who immigrated to the U.S.A. and crafted his skates with a few Dutch and American characteristics. If they could only talk. Length 12", Width 2", Platform height 2", Curl diameter 4 5/8", c. 1850. Class E: $350-500.

Figure 33. This pair of Dutch skates has hand forged runners and nice curls with turn-overs at the tips. The top of the runner has a decorative touch near the platform intersection. The platform shape is rectangular with straight sides and has three morticed slots for leather straps. The heel screw has four barbs to hold the boot secure. Length 15 5/8", Width 2 1/8", Platform height 1 1/2", Curl diameter 3 3/8", c. 1870. Class C: $100-200.

Figure 34. These skates have a very high, hand forged, handmade runner with a graceful curl. The platform is also very thick and has three morticed slots for the leather strap fastenings. The wooden platform was painted blue originally and traces are still evident. The heel has a hand wrought spike and the toe area has sharp nails preventing boot slippage. Note the curved bottom on the runner. Butterfly morticed notches on top of the runner provides a means to incorporate a screw of that shape on which to fasten. The skates certainly have the Dutch look, but could possibly be American made as well. Length 12", Width 2 5/8", Platform height 2 1/4", Curl diameter 4 5/8", c. 1850. Class E: $350-500.

Figure 35. A later transitional pair of Dutch skates that incorporate the attached leather uppers. The uppers are tightened with strings and the wood platforms have two traditional slots for leather straps. A brass trim piece is used on the heel for support and decoration. The skates were worn over the skater's boots. They were stamped and made by the L.K. Hoekstra & Co., Wargo, Holland. Length 11", Width 3", Platform height 1 3/4", c. 1920. Class C: $100-200.

Figure 36. A newer pair of Dutch skates that is known as the "Saw Skate." The bottom of the blade runner has two nicks or notches designed to bite the ice better and give the skater a faster start-up. Perhaps the design never caught on or proved successful and then fizzled out, like the American Ford automobile named the "Edsel." Both are now shown in museums! The skates are painted a dirty white color and are called Zaagschaats made by E. Vonk. Co., Oudeschoot, Friesland, Holland. Length 17 1/4", Width 2 3/8", Platform height 1 1/2", c. 1950. Class C: $100-200.

Figure 37. These Dutch skates incorporated the safety toe into the curl and had straight platform sides. There are two slots for leather straps in the yellow painted platforms. The heel has two barb nails on it for boot support. They were made by the Ijsnocht Co., Holland. Length 14", Width 1 3/4", Platform height 1 1/4", c. 1890. Class B: $25-100.

Figure 38. Dutch skates with nice brass finials at the tips of the safety curls. A brass heel cup is screwed to the rear of the platform. The platform heel also has the traditional four-pointed barbed nut, as well as three leather straps for fastening. Length 13 1/4", Width 1 7/8", Platform height 1 1/4", Curl height 3 3/8", c. 1880. Class D: $200-350.

Figure 39. Beautiful Dutch skates with brass acorn finials at the tips of the curls. The platform is made of dark walnut and is rectangular. It also has a three-barbed heel nut to secure the boot. The harness on the rear is hand stitched with linen thread and passes through the front slot for the fastening arrangement. Length 16", Width 2", Platform height 1 1/4", Curl height 3 3/4", c. 1860. Class E: $350-500.

Chapter Three

England

Elegant Skates & Victorian Figure Skating

Skating became very popular in the Lake Districts of England in the 1600s. Later, in the first half of the 1700s, the figure skate was developed in England. Its curved bottom runner allowed the skater to make sharp corners or close turns in small areas or on ponds, making a "figure" design on the ice. The heel of the runner was also curved up a bit so the skater could skate backwards in a variety of routines. Victorian or elegant skating on the high bladed curved figure skates became very popular in England. Accomplished skaters from around the world developed figure skating into an art. The flat bladed heel of the speed skate runner, on the other hand, would not allow skating backwards, as it would stub the irregular ice and cause an accident.

Since Sheffield, England has been known for many years as the manufacturing center for high quality steel tools, it was a natural for the tool steel makers there to manufacture and market ice skates. Ice skate blades were made of high quality steel. In the early years, all of the ice skates were hand crafted, until the demand required the companies to start making them in high volume quantities. The companies developed specialized departments to make each part of the skate in a production manner.

The skate makers in the Sheffield area of England include Marsden Brothers, James Howarth, Gray And Selby, William Marples and Sons, I. Sorby Co., and Hunter and Son. Marsden Brothers, has made high quality skates since 1696. *Figure* 40 illustrates two pairs of specially made skates for Queen Victoria and Prince Albert. They are called "Swan" skates with the swan's head on the front of the runner serving as the curl or prow.

MARSDEN BROS
SHEFFIELD.

Opposite page

Figure 40. These two spectacular pairs of Swan's ice skates were specially made for the British Royal family, Queen Victoria and Prince Albert, 1840. The makers were Marsden Brothers of Sheffield, England, who have been skate makers since 1696. Note the company stamps on the runner along with what appears to be a greyhound as their corporate mark. Prince Albert's skate platforms seem to be made out of rosewood. Decorative brass inlays on the toe plate and a star and circle on the heel plate can be seen. Sharp spikes on the platform and a screw on the heel were used to secure the boot. The runner blades seem to be made of silver with a fantastic curl or a swan's head design with engraving. The leathers appear to be patent leather with a soft liner inside. Queen Victoria's skates have patent leather uppers with decorative stitching. Silver metal skirts are screwed to the platform attaching the leathers. The heel cup appears to be made of silver and engraved. A leather ring harness attachment is also used to go up over the foot's arch. The craftsmanship in both pairs of skates is unbelievable! *Courtesy of the World Figure Skating Museum and Hall of Fame.*

Figure 41. A beautiful pair of English Victorian skates bears a star and scalloped brass inlay plates on the platform. The platforms are made of walnut with a heel screw and sharp nails to prevent boot slippage. The front runner has a nice long turn up curl. Original leather straps through three morticed slots were used for fastening. The craftsmanship in the English skates is outstanding. They were manufactured by the William Marples and Sons of Sheffield, England. This company has been well known internationally for making quality craft tools for many years. Length 13 1/2", Width 2 1/4", Platform height 1 1/2", Curl height 1 3/4", c. 1850. Class F: $500-750.

Figure 42. An elegant and exceptionally long pair of English racing skates. The platform has nicely decorative brass inlays at the toe, foot, and heel areas. The foot area has two sharp spikes mounted in the brass plate and the heel has a long screw mounted to prevent boot slippage. The platform is made of mahogany, which has a beautiful red patina finish. The fastenings consist of three leather straps. The long runner has a nice curl design on the front and is extended beyond the heel on the rear. This pair of skates is very attractive and well made. Length 16", Width 2 3/8", Platform height 1 1/2", c. 1860. Class F: $500-750.

Figure 43. This unusual pair of English skates has decoratively engraved brass foot and heel plates. The runner is riveted to two brass stanchions and has a nice turn-up prow on the front. Holes in the brass platforms were used to fasten straps by rivets. These unique skates were manufactured by the Gray and Selby Co., Nottingham, England. Length 12", Width 2 1/2", Platform height 1 1/4", Curl height 1 1/2", c. 1860. Class F: $500-750.

Figure 44. A very rare and finely made pair of English women's skates. They were made by the Marsden Brother's Skate Co. of Sheffield, England. The skate runner is stamped "Marsden Brother's Skate Manufacturers by special appointment to her Majesty and the Royal family, Sheffield." The manufacturer's corporate mark seems to be a greyhound. The government gave certain quality manufacturing companies permission to put this prestigious stamp of approval on their products. These skates are an example of that tradition. The platforms are made of walnut and the enclosed foot leathers are made of patent leather with the original strings still attached to tighten up the toe. The socketed heel cups are made of German silver. The runner has a very nice curl and knob finial. These were purchased at a London auction. Length 12", Width 2 1/2", Platform height 1 1/2", Curl diameter 2 1/4", c. 1860. Class F: $500-750.

Figure 45. An attractive pair of women's English skates with nicely forged turn-up curls on the fronts of the runners. The runner is stamped with a shield enclosing an X. The platforms are made from tiger stripe maple with 70% of the original varnish remaining. Three slots were used for leather fastenings. Length 11 5/8", Width 1 7/8", Platform height 1 1/4", Curl height 1 3/4", c. 1850. Class E: $350-500.

Figure 46. A pair of English figure skates manufactured by Colouhoun and Cadmam Skate Co., Sheffield, England. The manufacturer's stamp is visible on the runner. The beech platforms have a brass inlay plate at the toe with sharp spikes and heel screws to secure the boot. Wide enclosed straps were used at the toe and regular straps at the heel for fastenings. Length 10 1/2", Width 2 1/4", Platform height 1 1/2", c. 1870. Class D: $200-350.

Figure 47. English figure skates manufactured by the I. Sorby Co. of Sheffield, England. The runner blade is stamped with a company logo, a man in Victorian dress and hat with his right hand raised. The platform is made from walnut and still has 95% of its original varnish finish. Two sharp spikes are located on the foot pad and a screw on the heel to secure the boot. Two leathers were used for fastening. Length 10 5/8", Width 2 1/8", Platform height 1 1/2", c. 1870. Class D: $200-350.

Figure 48. A pair of English figure skates with runner blades that have nice round noses on both ends. The platform is contoured to the foot and has a step down for the heel. The top of the platform has two decorative brass inlaid plates on either end. The heel plate is an unusual design and is eye catching with the brass fastening nut. The platform has three sharp spikes on the foot plate and a long screw on the heel for boot support. The toe enclosure and strap on the back is used for fastenings. Length 12", Width 2 1/2", Platform height 1 1/2", c. 1870. Class D: $200-350.

Figure 49. English figure skates with unusual square-shaped toes. The runners are stamped cast steel and have curved blades on both ends. The bottom of the runner blade has a deep hollow or gutter cut in it. The platforms have brass plate inlays on both front and rear. The heel has a long sharp spike for securing the boot. The toe front had a wide leather enclosure and rear straps for fastening. Length 10 1/2", Width 2 1/8", Platform height 1 3/4", c. 1860. Class D: $200-350.

Figure 50. This pair of English figure skates has brass inlaid plates on the toes and heels. The platform is made of walnut with three slots for leather straps. The platform has spikes in the toe area and a metal pin on the heel to fit into a hole in the boot heel. Length 11", Width 2 1/2", Platform height 2", c. 1860. Class D: $200-350.

Figure 51. These skates have rosewood platforms with a contoured profile on top. The platform is fastened to the runner with a slotted nut on the front and a screw on the rear. The runners are very thick and heavy, 1/4" cast steel. Two leathers were used to fasten the skates to the boot. They were made by the Williams and Co. and have that English look. Length 11 1/4", Width 2 1/4", Platform height 2", c. 1870. Class D: $200-350.

Figure 52. A pair of English skates manufactured by the John Wilson Co., Sheffield, England. The manufacturer's stamp on the runner states "The Mount Charles, Dowler's Patent Portland Works Sheffield." The corporate mark of a greyhound was carried over since the John Wilson Co. bought out the Marsden Brother's Co., which previously used the mark. Length 12", Width 2 1/4", Platform height 1 3/8", c. 1890. Class C: $100-200.

Chapter Four

Germany

Large Curls & Quality Exporters

The Remscheid area of Germany was known for its cutlery and tool steel making for years. It was a coal mining area with the natural resources available to make quality steel. It became the center of ice skate manufacturing in Germany and eventually developed into one of the skate capitals of the world. The skate blades required good, hard steel and it was a natural product for the Remscheid area tool makers. They produced high quality skates at low cost and the area became one of the major exporters of skates all around the world, especially to countries like Holland, England, and America.

On August 12, 1833 the C.W. Wirths Company (Carl Wilhelm Wirths) of Hutz, Germany exported 600 pair of high quality ice skates to the Christian Hesser Co. of Philadelphia, Pennsylvania. (surviving bill of sale, courtesy of Smithsonian Institution). These skates had an exceptionally large curl on the front of the blade with a decorative brass finial at the tip as shown in *Figure 53*.

A few of the early skate makers in the Remscheid area of Germany are C.W. Wirths, Eduard Engels, Arnold Goldenberg, Peter Janson, Peter Tillmanns, and Milger and Sons.

Figure 54. A cold winter scene in Ohio Amish country, a German settlement that still practices the old traditions of farming and hand craftsmanship. This view illustrates the common practice of using the horse and buggy for transportation. The Amish were gathered across the road for a farm auction. The blacksmith, carpenter, and harness maker shops still exist in the German Amish community, trades that would have fashioned the old ice skates during the 19th century.

Figure 55. A nice pair of German skates with large curls and brass acorn finials at the tips. The platform has four slots to accommodate leather straps for fastening. The platform has a screw mounted in the heel area to secure the boot. Length 2 1/2", Width 2 1/2", Platform height 2", c. 1840. Class F: $500-750.

Opposite page

Figure 53. Skates with exceptionally large curled, hand forged runners with brass acorns on the tips. Note the brass tips are curled around and heading downward. They were made by the C.W. Wirth Co., of Hutz, Germany. This is in the Remscheid area, a center of German craftsmanship for tools and ice skates. The platforms are secured to the runner by a clamp barb and metal pin. The rear of the platform has a cut down step to accommodate the heels. Handmade screws were inserted in the heel to prevent boot slippage. Three leather straps were used to fasten the skates to the boot. Length 12 5/8", Width 2 1/2", Platform height 1 3/4", c. 1830. Class F: $500-750.

Figure 56. This pair of skates has exceptionally large curls on the fronts of the runners and brass finials at the tips. They were made by the Hilger & Son's Co., of Remscheid, Germany. The wooden platforms were made of tiger stripe maple with a cut down heel area. Three leather straps were used for fastening. A screw is in the heel to prevent boot slippage. Length 13 1/8", Width 2 1/2", Platform height 1 3/4", Curl diameter 5 5/8", c. 1830. Class F: $500-750.

Figure 58. An exceptional pair of ice skates made by David Wirths of Holterfeld, Germany. They have beautiful heel cups made of cast brass. The large curls on the runners have unusual brass balls for finial tips. There were three leather straps for fastening to the boots. The bold stamped name of D. Wirths on the forged runner has the "D" turned counterclockwise 90 degrees. This pair of skates is very attractive and would be considered a keeper. Length 14", Width 2 1/2", Platform height 1 3/4", Curl Diameter 5 1/8", c. 1840. Class F: $500-750.

Figure 57. Red painted platform skates with large turn up curls and brass acorn tips. The runners below the platform were also painted red. Three leather straps were used on the platform. A handmade spike is used on the heel and metal barbs on the foot pad to hold the boot. Length 12 3/8", Width 2", Platform height 1 3/4", Curl diameter 4 1/2", c. 1850. Class E: $350-500.

Figure 60. German made skates perched on top of a wheat shock in Amish country. The runners are very fancy with a decorative design on top near the platforms. They have round finials at the tips of the curls. The platform plan view is sawed out in the form of an hourglass. It has a heel screw with four barbs to secure the heel from slipping. The skate design is Dutch and is called a Breinermoor model. It was a popular skate style that was exported into the Breinermoor district of Friesland, Holland, from Germany. Length 15 3/4", Width 2 1/2", Platform height 1 3/8", Tip height 1 7/8", c. 1870. Class D: $200-350.

Figure 59. A pair of skates with exceptionally large curls and brass acorn finials at the tips. The heel has screws and the toe has metal barbs to prevent slippage. This photo was taken on top of a fence post with wheat shocks in the background. The shocks are in a German Amish community located in Holmes County, southeast Ohio. Length 13 5/8", Width 2 3/8", Platform height 1 3/4", Curl diameter 5 1/4", c. 1850. Class F: $500-750.

Figure 62. A Breinermoor model skate with a differently shaped platform. It has the typical decorative runner with the scalloped top near the platform. The graceful turn up curl has a nice onion dome round finial. The heel screw has four barbs to secure the boot to the platform. Two leather straps are used to hold the boot secure. The platform has some decorative carving on it. Length 16 1/2", Width 2 1/2", Platform height 1 5/8", Tip height 2 1/2", c. 1870. Class D: $200-350.

Figure 61. German skates with very decorative fronts. The top of the runner has a decorative scalloped design and fancy turned finial at the tip. The platform has two leather straps to fasten the boot. The heel has four barbs to secure the boot. This is a different design for a Breinermoor skate. Length 14 1/2", Width 2 1/8", Platform height 1 1/8", Tip height 1 7/8", c. 1870. Class D: $200-350.

Figure 63. A handmade pair of skates given to me by a friend in northern Germany. His father made the steel runners from an old saw blade, which had good steel in it. The small curled runner has the safety toe feature with the runner sandwiched between the wooden platform. There are three morticed slots in the platform for leather strap fastenings and a handmade spike at the heel to prevent boot slippage. The skates certainly have the Dutch influence and look. Note the German Konig pilsner beer next to the skates, a good local beer. Length 12", Width 2", Platform height 1 5/8", Curl height 2", c. 1925. Class C: $100-200.

Opposite page

Figure 65. A very finely made pair of skates made by the Wirths Brothers of Hutz, Germany, Remscheid area. The runners are beautifully hand forged into a large curl with large brass ball finials at the tips. They are also hollow ground. The company mark stamped on the blade reads BRS Wirths with an animal logo and touch marks on either side. The platforms have three leather straps and a sharp nail on the heel attached with a brass screw. The wooden platform has a nice patina and color. This is an outstanding pair of skates. Length 13 3/4", Width 2 1/4", Platform height 1 5/8", c. 1840. Class F: $500-750.

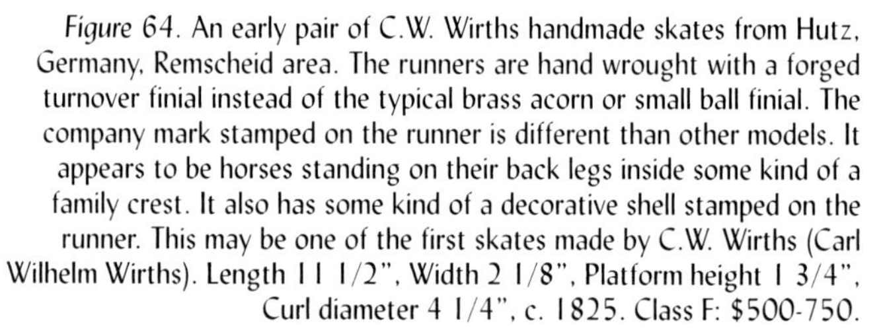

Figure 64. An early pair of C.W. Wirths handmade skates from Hutz, Germany, Remscheid area. The runners are hand wrought with a forged turnover finial instead of the typical brass acorn or small ball finial. The company mark stamped on the runner is different than other models. It appears to be horses standing on their back legs inside some kind of a family crest. It also has some kind of a decorative shell stamped on the runner. This may be one of the first skates made by C.W. Wirths (Carl Wilhelm Wirths). Length 11 1/2", Width 2 1/8", Platform height 1 3/4", Curl diameter 4 1/4", c. 1825. Class F: $500-750.

Chapter Five

America

The Innovators of New Skate Designs

Ice skating played an important part in the early development of our country. In America ice skating was used as a means of transportation, for hunting, and as a winter sport activity. The winters are not as cold today as they were in the 1700s and 1800s, so we do not get the long deep freezes needed for good outdoor skating. Not only were winters colder, the entire country had unlimited places to skate including the entire eastern coastline, lakes, rivers, ponds, and ditches. As the pioneers moved west in the early 1800s no community was ever in want of a place to skate during the winter months. Skating became a community, social, and family entertainment activity. Long distance skating, speed skate racing, and elegant figure skating competitions sprang up all over the country.

The earliest development of ice skating in America occurred in the eastern New England States. Cities like Philadelphia, New York, and Boston were the centers of early ice skating development and activity. Ice skating became the craze of the country in the 1860s as a family winter sport. In my estimation ice skating activity peaked in America in 1865.

Originally ice skating was predominantly a male sport. But in the late 1700s a few female skaters came onto the scene, and more in the 1800s when the social skate clubs were first formed. The first skating club in the world was reportedly formed in Edinburgh, Scotland in 1744.

In the U.S., ice skating clubs began to spring up in the larger eastern cities. The oldest skating club in America was the Philadelphia Skating Club founded on December 21, 1849. Other skating clubs that formed in the east were the New York Skating Club, 1863, Cambridge Skating Club, Brooklyn Skating Rink Association, and the Boston Skating Club. The National Skating Association was founded in the year 1879. *Figure 66* is a photograph of the New York Skating Club with its members all dressed up standing on the dock.

Figure 66. The old New York Skating Club on *The Fifth Avenue Pond* 1863. Note the two old wooden wheel bicycles known as "boneshakers" on the dock. The people all seem to be dressed up in their finest clothes including top hats. The children are also dressed "for church." It must have been a men's club judging from the absence of women.

The Smithsonian Institute's booklet titled, "The American Skating Mania," states that between 1863 and 1867, nine skating clubs were established in Philadelphia. "By 1868 indoor skating rinks had been built not only in Philadelphia, but in Chicago, St. Louis, Cincinnati, Cleveland, Pittsburgh, Quincy and Springfield, Ill., Indianapolis, Columbus, Buffalo, Rochester, Syracuse, Oswego, Boston, Brooklyn, and Jersey City."

As the sport of ice skating became popular, a few individuals became world celebrities. One such well-known skater was Jackson Haines who became a champion in America and was known as "The American skating king." He became one of the world's greatest professional skaters, giving exhibitions in America as well as Europe. He established the style known as "fancy figure skating" around 1865 and is credited as being the founder of the "international style of skating". He invented the first steel toe and heel plates forged to the runner, which were then screwed directly to the heel and sole of the shoe. Haines was also the first to add teeth to the front of the blade.

Sonja Henie, a Norwegian skater, was also well known in America. She was a world champion in the international ice skating competition scene of the 1920s and 1930s. She won several Norwegian, European, world, and Olympic gold medals in competitive figure skating. She toured America in several ice skating exhibitions and starred in many Hollywood films. She was certainly influential in the further development and promotion of figure skating worldwide. She revolutionized figure skating and skate clothing styles for women internationally.

The painters and lithographers also got onto the band wagon of ice skating in America by capturing on canvas the activities of ice skaters on the lakes and ponds. *Figure* 67 shows a Currier and Ives print titled, "The Skating Pond," depicting Central Park in the winter of 1863. *Figure* 68 is a lithograph by Nathaniel Curriers titled, "Winter Pastime," which captures a rural scene in 1856. *Figure* 69 illustrates a river scene of skaters with the curled front runners on their skates and fashionable winter clothing. In the foreground one of the men is on his back and another is falling down. The origin of the print and artist is unknown. I purchased it mounted on a wooden board several years ago. A fourth print of a winter scene (*Figure* 70) is by Louis Prang titled, "Skating 1885." Interestingly, Louis Prang, a German immigrant, is known as the father of the American Christmas card. He was a lithographer and artist who painted and printed the first Christmas cards in America. *Figure* 71 is a wood cut from the February 10,1866 issue of Harper's Weekly illustrating the "Great Skating Rink at Chicago." The people seem to be doing a formal ice skating dance. Note the brass instruments on stage with the horns sticking up over the musician's shoulders like the Civil War band horns.

Figure 67. A Currier and Ives print titled *The Skating Pond in Central Park, Winter of 1863*. Note all the skaters dressed up in their colorful Victorian costumes. Several of the men are wearing formal top hats. A few have fallen and are sitting on the ice wondering if they should give it another try.

Figure 68. A lithograph by Nathaniel Currier titled *Winter Pastime*, which captures a rural scene in 1856. Everyone seems to be enjoying the winter afternoon. Boys are playing ice hockey on the pond while others are sledding down the hill. The older couple is riding down the lane in a horse drawn sleigh. It appears that the boys are coaxing the young lady into playing hockey with them. One of the boys is retightening his skates.

Figure 69. This river scene shows several skaters with curled runners on their skates and fashionable winter clothing. In the foreground one of the men is on his back and another is falling down onto him. All of them seem to have on their best Sunday duds. It looks like a wide river and big bridge are in the background.

Figure 70. A colorful winter scene of ice skaters enjoying themselves. It is a chromolithograph by Louis Prang, titled *Skating*, 1885, Boston. Note the elegant Victorian clothing worn by the skaters. On the right a skater is on his back and another on his way. *Courtesy of The Library of Congress, Washington, D.C.*

Figure 71. A wood cut from the February 10, 1866 issue of *Harper's Weekly* illustrating the Great Skating Rink at Chicago. The skaters seem to be doing a formal ice skating dance. Note the brass instruments on stage with the horns sticking up over the musician's shoulders, like the horns used in the bands during the civil war.

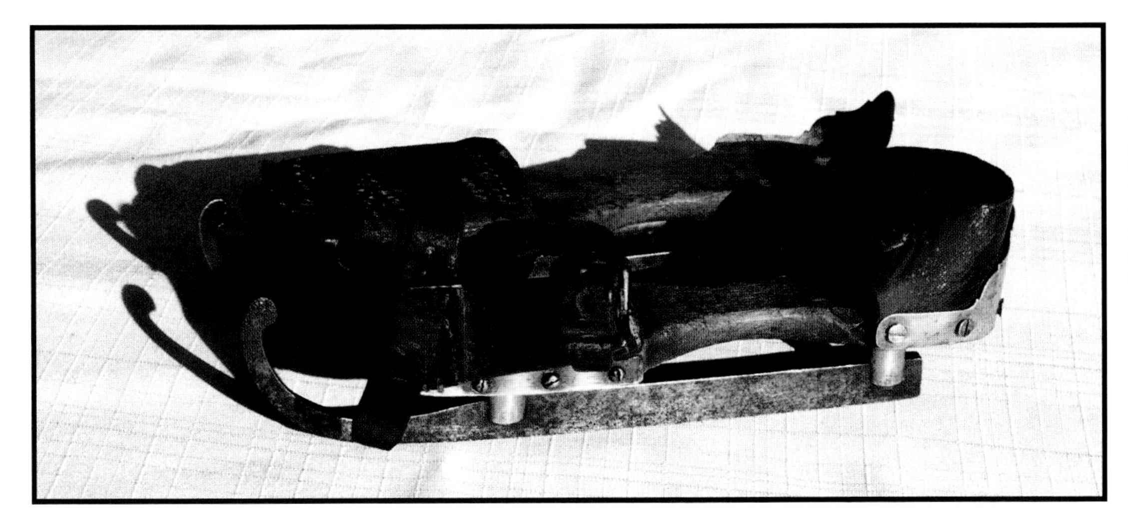

Figure 72. A pair of skates manufactured by the Fredrick Stevens Co., the first skate manufacturer in the United States, founded in 1839. The runner has a nice little curl and tip design on the end. The stanchions are made of brass and fastened to the runners with pins. The heel has a decorative brass support holding the leather cup in place. The front leathers cover the entire toe area held on with a brass trim strap. Length 9 1/2", Width 2 3/8", Platform height 1 5/8", c. 1840. Class D: $200-350.

In the eighteenth century most ice skates were fashioned by hand by blacksmiths and woodworkers, for the rural community and the city folks alike. But in the nineteenth century all this interest in ice skating in America created a big demand for ice skates; several thousands of pairs were needed. Some were imported from Holland, Germany, England, and Sweden, but new ice skate manufacturers started to sprout up around America. The first ice skate manufacturer in America was the Fredrick Stevens Co. founded in 1839. *See Figure* 72.

Several other ice skate manufacturers started up primarily in the New England states in the mid-1800s to meet the demand. They made a large variety of designs and styles over the years. The following names are but a few of the better known skate making companies in the business at that time.

Fredrick Stevens Co.	New York, NY	1839
Union Hardware Co.	Torrington, CT	1854
Samuel Winslow Skate Co.	Worcester, MA	1856
William Hawkins Co.	Birmingham, CT	1859
H. Clark Co.	Jordon, NY	1861
Douglas Rogers and Co.	Norwich, CT	1862
Barney and Berry Co.	Springfield, MA	1865
Keen Mfg. Co.	Keene, NH	1888

Americans were very inventive in the mid-1800s and submitted several skate designs for patents to the U.S. Patent Office in Washington D.C. They were to "improve" the looks and more importantly to create new designs to prevent skates from coming loose or off the shoes. See the patent section later which illustrates several innovative designs.

Union Hardware Company, Torrington, Connecticut and other American Skate Manufacturers

The second oldest and one of the largest ice skate manufactures in the country was the Union Hardware Co. in Torrington, Conn. They started business in 1854 and bought out Fredrick Stevens Co. on Feb. 13, 1872. The Union Hardware Co. eventually became part of the Brunswick Corp. in 1960 when they expanded and diversified into other sporting good lines. They continued making ice skates up until 1967 when they were phased out.

According to a Torrington newspaper article dated Oct. 24, 1967 the Brunswick Corp. (parent Co. of Union) announced on Oct. 19, 1967 that they had discontinued making ice skates as one of their product lines. It further stated, "Mr. George D. Wadhams is credited with the birth of the ice skating business in Torrington, an industry later assumed by the old Union Hardware Co., which made the name Torrington known throughout the world." They made several different skate designs throughout their history, a few of which are shown in *Figures* 74-81. They are from a catalog dated 1881. Notice the beautiful hand drawn artwork on the skates in the catalog.

The company made early wooden platform skates as well as the metal club skate designs, as they were then called and known. A Union Hardware Co. Club Skate can be readily identified by its logo or symbol, a cloverleaf design usually stamped out of the metal foot plate as illustrated in *Figure* 73. This is one of the more popular skates made by Union Hardware Co.

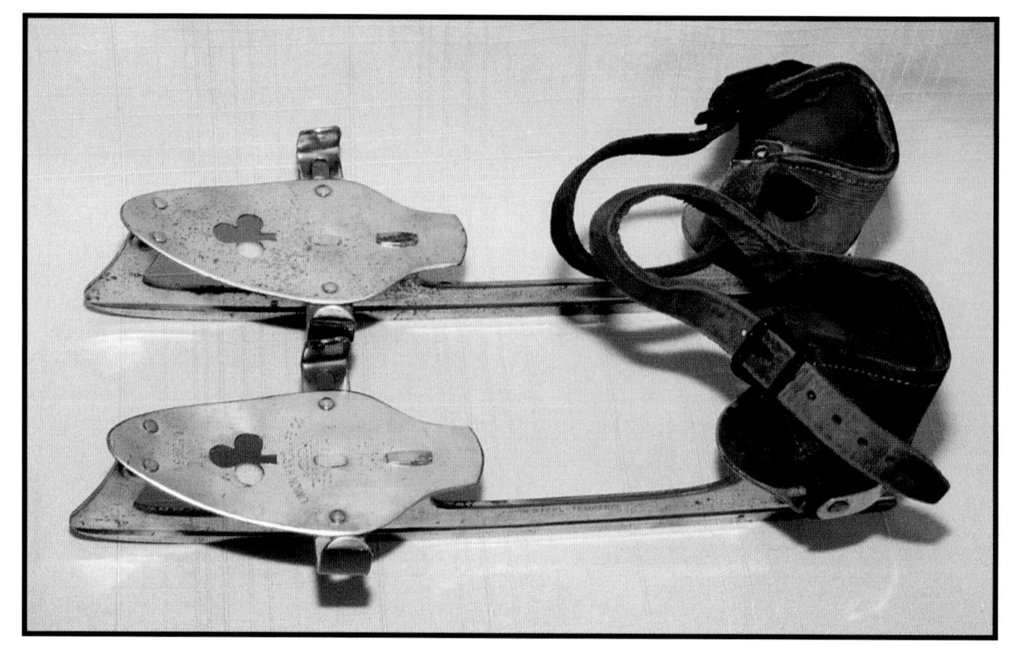

Figure 73. The Union Hardware Co. club skate model No. 5924 1/2 with leather cup heel straps. It is nickel plated with the famous company cloverleaf logo stamped out of the foot plate. The foot plate has adjustable clips that tighten up on the boot and hold it in place. Length 12", Width 2 1/2", Platform height 1 3/4", c. 1880. Class B: $25-100.

Union Hardware Co.

DOUBLE CLAMP SKATE.

Patented May 30, 1876; December 10, 1878.

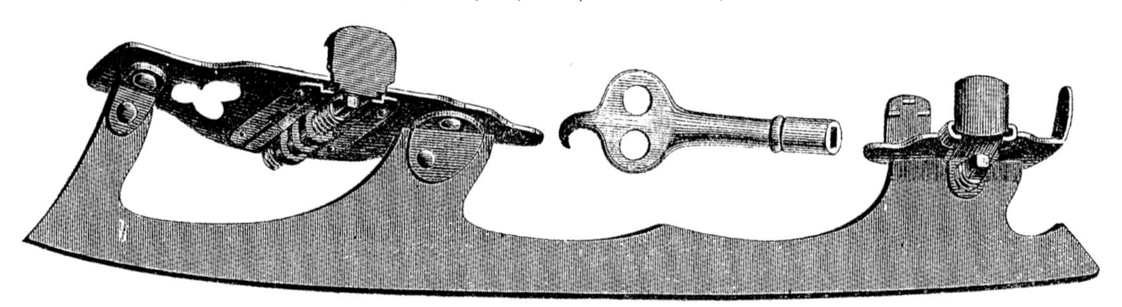

No. 6.

Sizes—8, 8½, 9, 9½, 10, 10½, 11, 11½ inches.

Double Clamp Skate, made with crucible steel sole and heel plates, provided with clamps securely fastened with our patent steel guides; operated with the best gun metal, right and left screw collared in the centre to give equal lateral motion to each clamp.

The runners are made of the best steel and iron, thoroughly welded and tempered in the best manner, and highly polished.

No means of fastening surpasses this for strength, security and durability.

No. 6, Blue Top, price per pair, . . $3.50. | No. 7, Nickel Top, price per pair, . . $4.50.

Figure 74. Union Hardware Co. double clamp skate No. 6 made out of steel; from the 1881 catalog. *Courtesy of The Torrington Historical Society, Inc.*

Union Hardware Co.

NEW YORK CLUB SKATE.

Patented May 30, 1876; December 10, 1878.

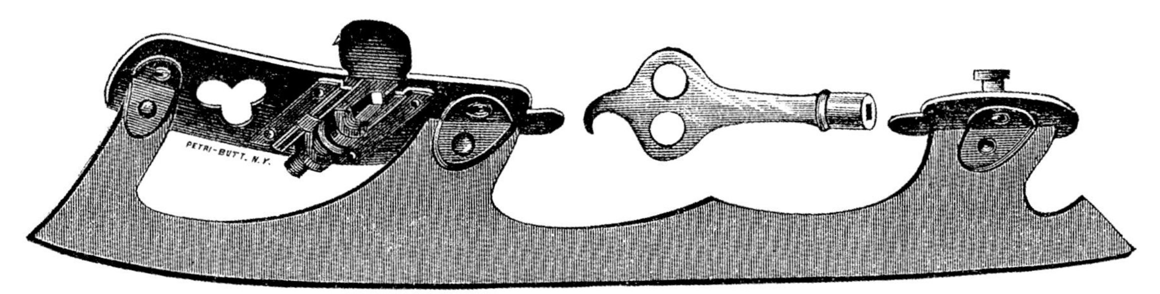

No. 4.

Sizes—8, 8½, 9, 9½, 10, 10½, 11, 11½ inches.

Our best NEW YORK CLUB SKATE so favorably known for many years.

The runners are made of the best iron and steel, thoroughly welded, tempered in the best manner, and highly polished. Clamps operated with best quality right and left screw, securely fastened with patent guide that holds the screw in the centre to give equal lateral motion to each clamp. Crucible steel foot and heel plates.

Warranted first quality in every respect.

No. 4, Blue Top, price per pair, . $3.00. | No. 5, Polished and Nickel Plated, price per pair, . $4.00.

Figure 75. Union Hardware Co. New York club skate No. 4 made out of steel; from the 1881 catalog. *Courtesy of The Torrington Historical Society, Inc.*

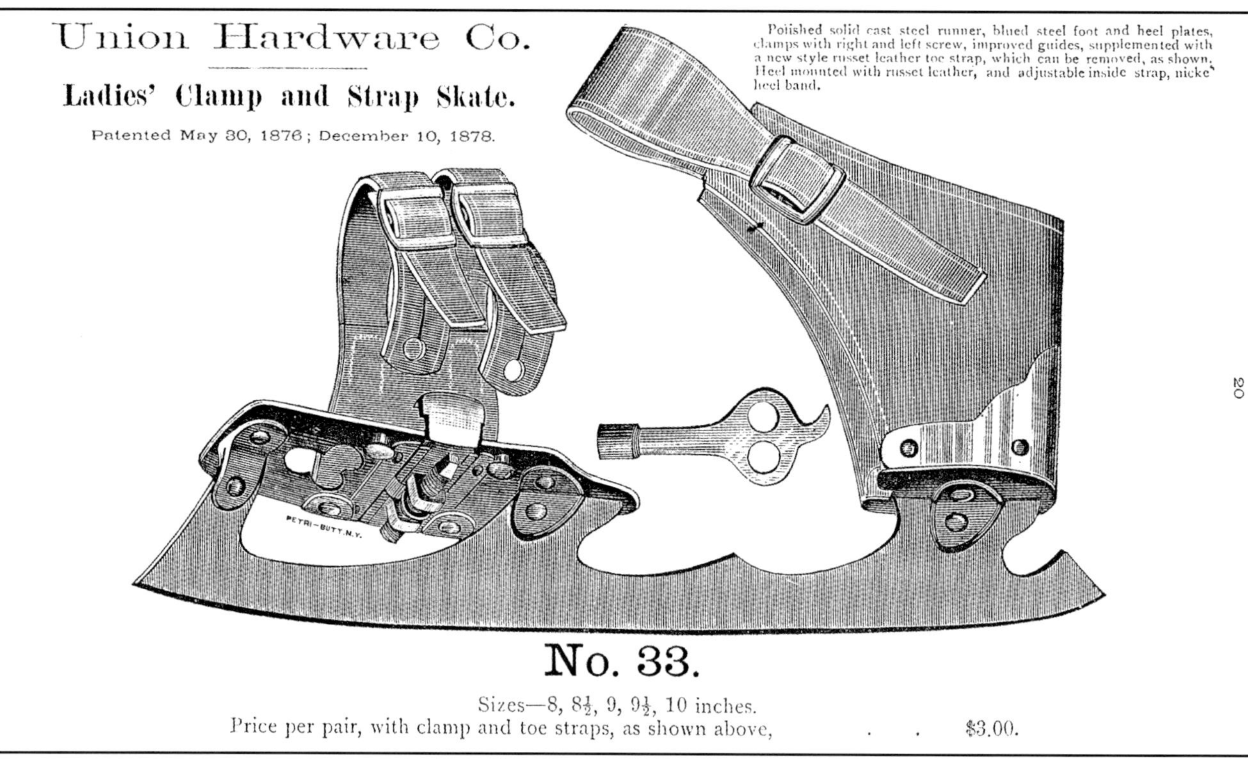

Figure 76. Union Hardware Co. ladies clamp and strap skate No. 33 made of steel with leather fastenings; from the 1881 catalog. *Courtesy of The Torrington Historical Society, Inc.*

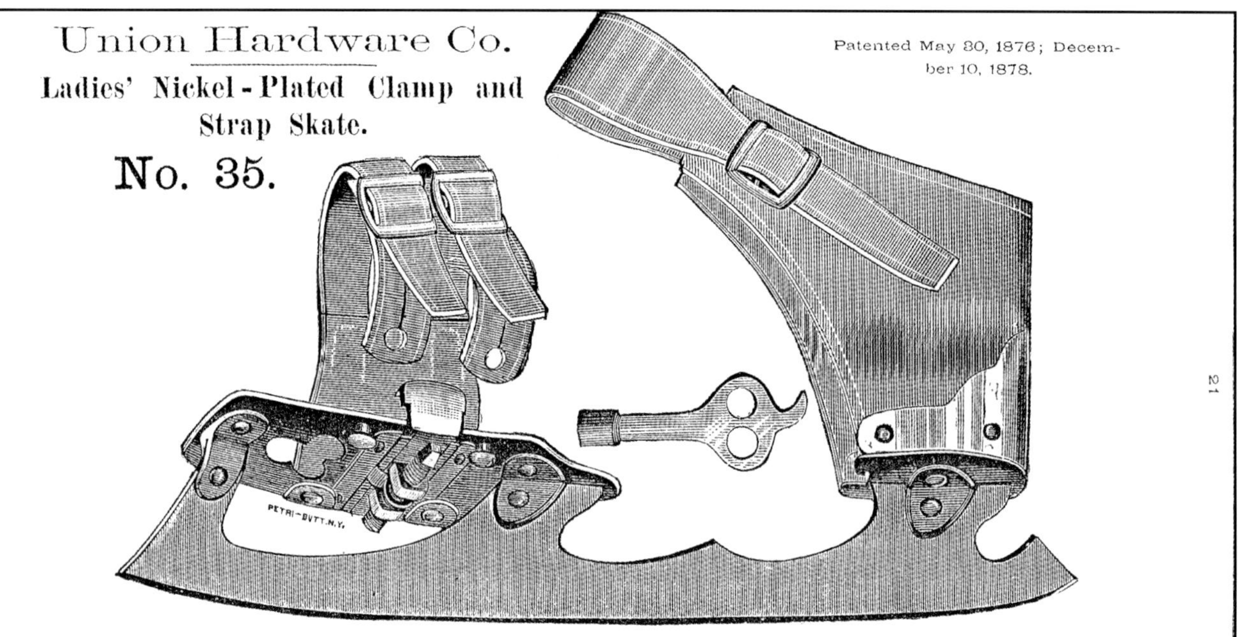

Figure 77. Union Hardware Co. nickel plated clamp and strap skate No. 35 made of steel with leather fastenings; from the 1881 catalog. *Courtesy of The Torrington Historical Society, Inc.*

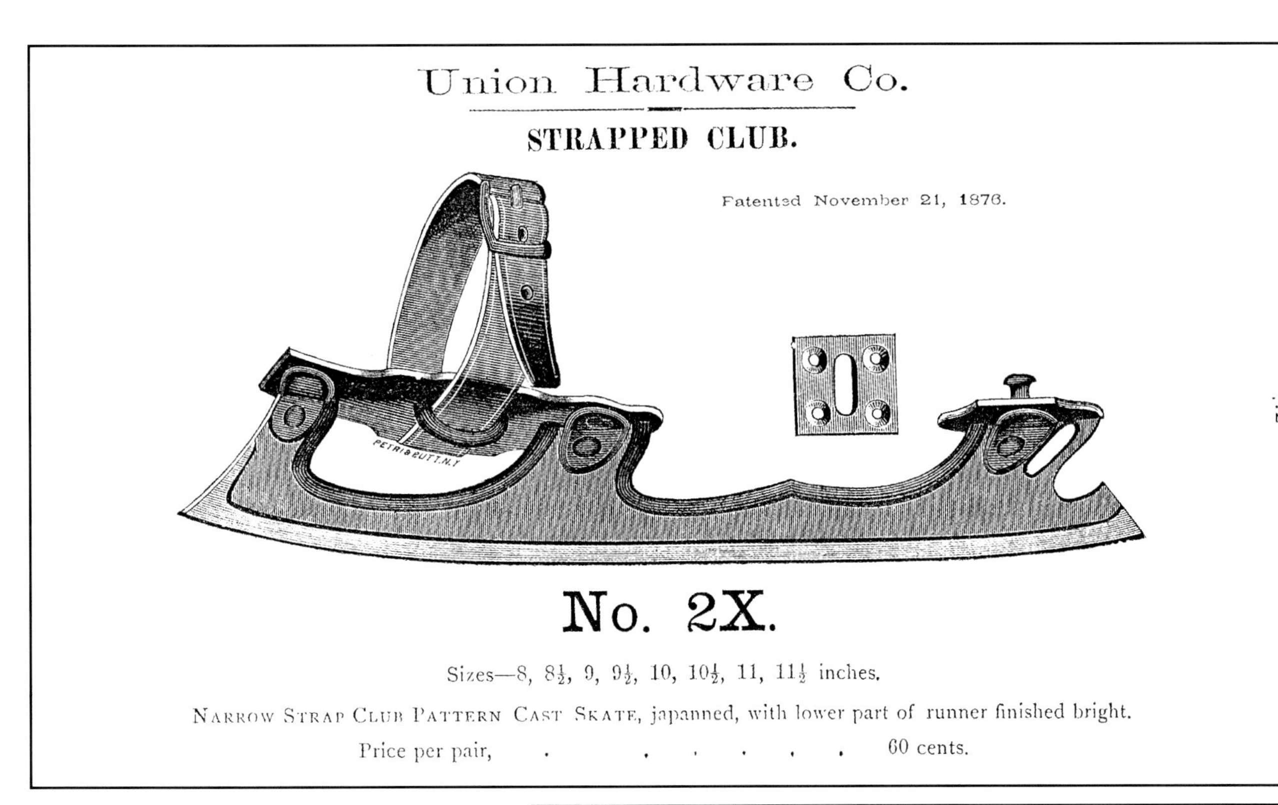
Union Hardware Co.

STRAPPED CLUB.

Patented November 21, 1876.

No. 2X.

Sizes—8, 8½, 9, 9½, 10, 10½, 11, 11½ inches.

NARROW STRAP CLUB PATTERN CAST SKATE, japanned, with lower part of runner finished bright.

Price per pair, ... 60 cents.

12

Figure 78. Union Hardware Co. strapped club skate made of cast steel with a leather toe strap No. 2x; from the 1881 catalog. *Courtesy of The Torrington Historical Society, Inc.*

Union Hardware Co.

GENTLEMEN'S SKATES.

No. 20. Half Rocker.

Sizes—7, 7½, 8, 8½, 9, 9½, 10, 10½, 11 inches.

Monitor pattern, iron frame skate, polished beech wood, with solid heel screws, pierced for short English straps.

Price per pair, ... 60 cents.

13

Figure 79. Union Hardware Co. gentlemen's skates No. 20, half rocker made of wood and steel; from the 1881 catalog. *Courtesy of The Torrington Historical Society, Inc.*

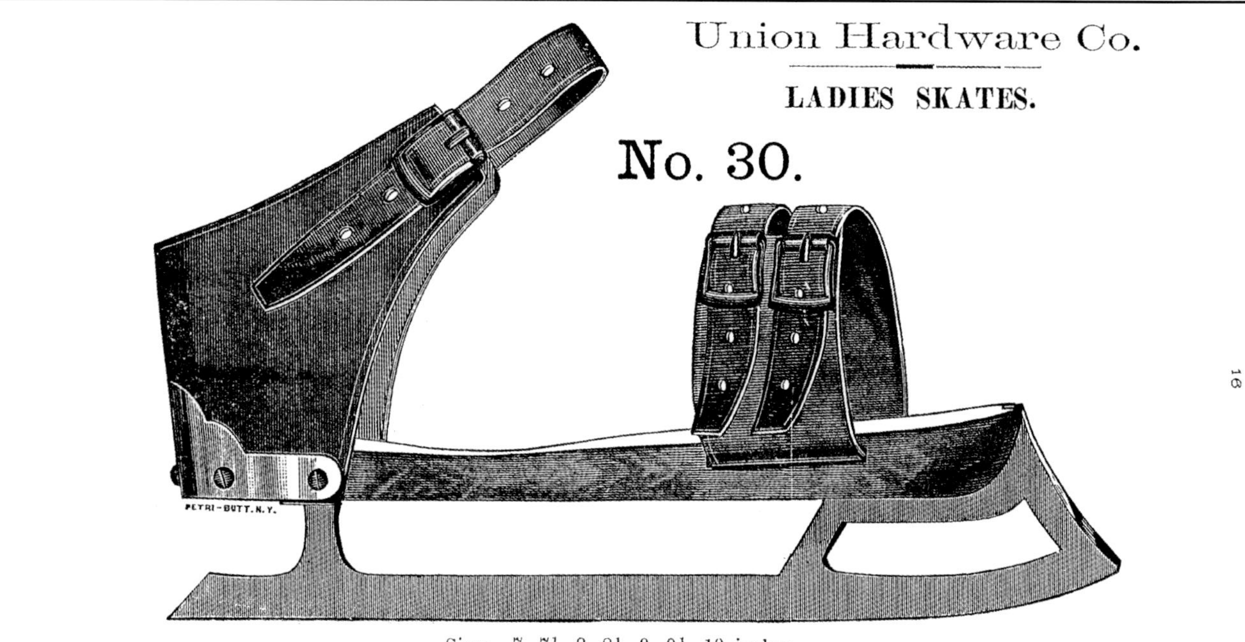

Figure 80. Union Hardware Co. ladies' skates No. 30 made of wood and steel with leather fastenings; from the 1881 catalog. *Courtesy of The Torrington Historical Society, Inc.*

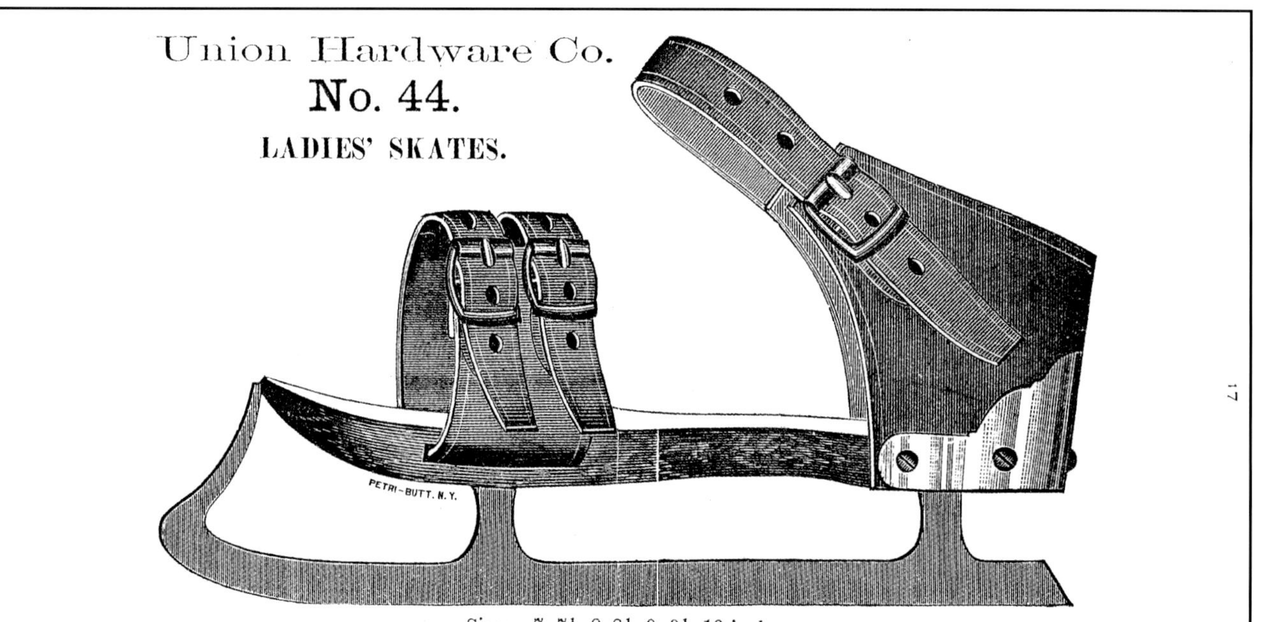

Figure 81. Union Hardware Co. ladies' skates No. 44 made of steel, wood, and leather fastenings; from the 1881 catalog. The runner design and platform front is a bit different. *Courtesy of The Torrington Historical Society, Inc.*

All Claims for Deductions from this Bill must be made within Ten Days after receipt of Goods.

WOLCOTTVILLE, CONN., Dec 3 1886

Mr Hiram Perkins

Bought of UNION HARDWARE CO.,

MANUFACTURERS OF

SKATES, SKATE WOODS AND SKATE STRAPS,

Screws, and all kinds of Skate Fixtures.

Terms,

	No.	8	8½	9	9½	10	10½	11 in.	Pairs.	Pr. Pair.	
Skates	40	3	3	3	3	3	1		16	75	12.00
	20	3		3		3			9	56	5.04
	80					1					3.50
Rocker	80					1					3.75
	½ Dz Lee Bond Straps								8.		1.33
											25.62
						15% off					3.84
											21.78
											3.40
											25.18

Paid

Figure 82. A copy of an old Union Hardware bill of sale dated Dec. 3, 1886. It describes several pairs of skates and their prices, including the skate number and size. Courtesy of The Torrington Historical Society, Inc.

An early bill from the Union Hardware Co. dated Dec. 3, 1886 for several pairs of skates is illustrated in Figure 82. Note the first item, 16 pair of skates at $.75 per pair, for a total of $12.00. I will take a couple dozen pair at that price!

The Union Hardware Company also made the popular wooden roller skates. Other items made included ice creepers, ice hatchets, bits, braces, chisels, draw knifes, hammers, dividers, handles, and cutlery to mention a few. The Union Hardware Co. made their mark in the ice skating world. Whenever a pair of metal ice skates are picked up at an antique market, the odds are great that the name Union Hardware will be stamped on them. They were both popular and plentiful in the American ice skating circles.

Another well known American manufacturer was the Samuel Winslow Skate Co. An original bill of sale dated December 17, 1890 is shown in *Figure 83*. The bill features 253 pair of skates of seven different styles for a total of $175.75, which averages $.69 a pair! That was a decent order going to a Troy, N.Y. firm in 1890.

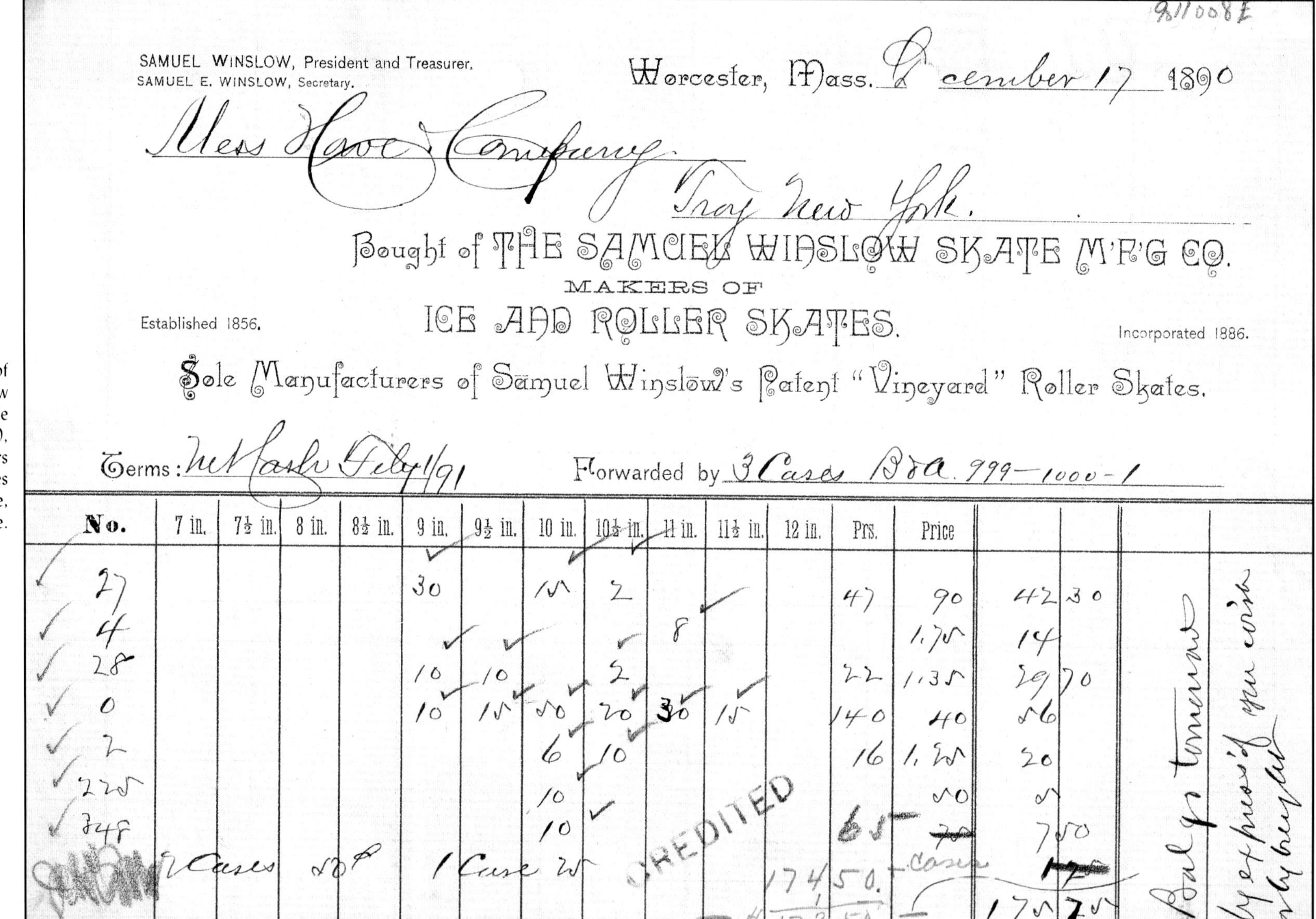

9110087

SAMUEL WINSLOW, President and Treasurer.
SAMUEL E. WINSLOW, Secretary.

Worcester, Mass. December 17 1890

Mess Howe Company
Troy New York.

Bought of THE SAMUEL WINSLOW SKATE M'F'G CO.
MAKERS OF
ICE AND ROLLER SKATES.

Established 1856. Incorporated 1886.

Sole Manufacturers of Samuel Winslow's Patent "Vineyard" Roller Skates.

Terms: Net Cash Feby 1/91 Forwarded by 3 Cases B&A. 999—1000—1

No.	7 in.	7½ in.	8 in.	8½ in.	9 in.	9½ in.	10 in.	10½ in.	11 in.	11½ in.	12 in.	Prs.	Price		
27					30		15	2				47	90	42	30
4									8				1.75	14	
28					10	10		2				22	1.35	29	70
0					10	15	50	20	30	15		140	40	56	
2							6	10				16	1.25	20	
225							10						50	5	
348							10					65	70	7	50
	2 Cases 50¢				1 Case 25							174 50	cases	175	75

CREDITED

Figure 83. A copy of an old Samuel Winslow Skate Co. bill of sale dated Dec. 17, 1890, showing several pairs of skates. It illustrates the skate number, size, and price.

The Spalding Company was another firm that marketed and sold ice skates, boots and shoes along with their other sporting goods lines. *Figures 84-86* illustrate a few of the skate designs and skating shoes.

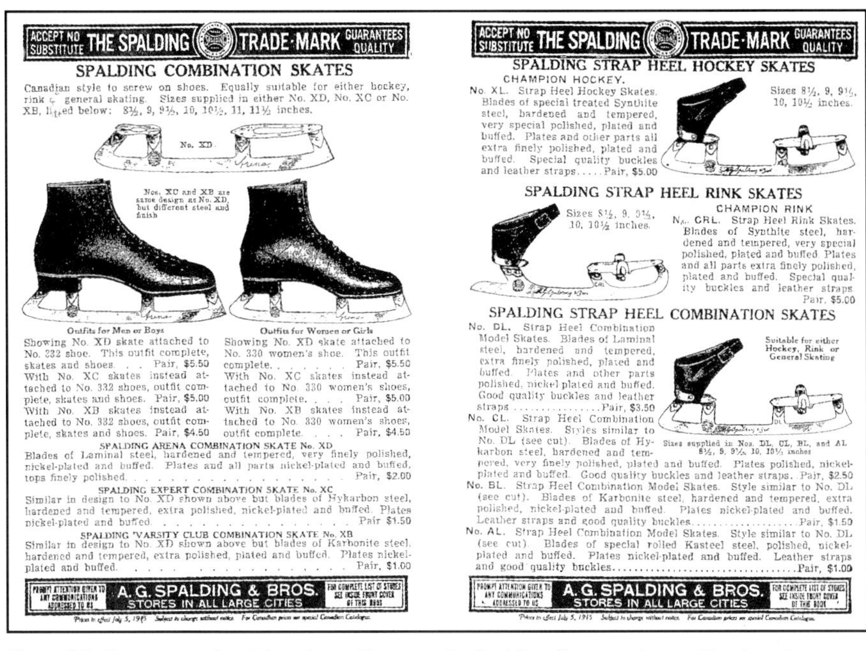

Figure 84. These two advertising sheets illustrate the Spalding Company's combination skates and strap heel hockey skates.

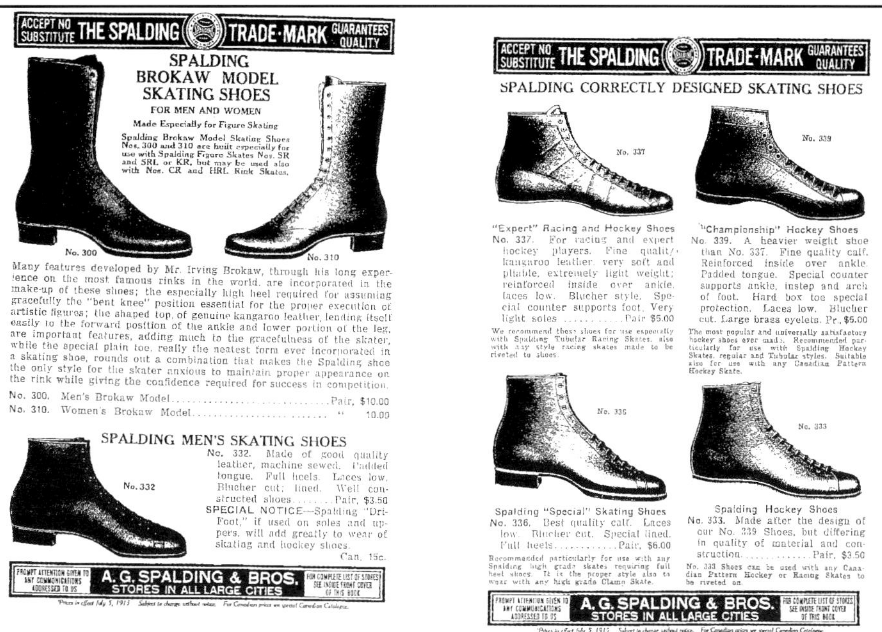

Figure 86. The Spalding Company's Brokow model skating shoes and correctly designed skating shoes are illustrated on these two advertising sheets.

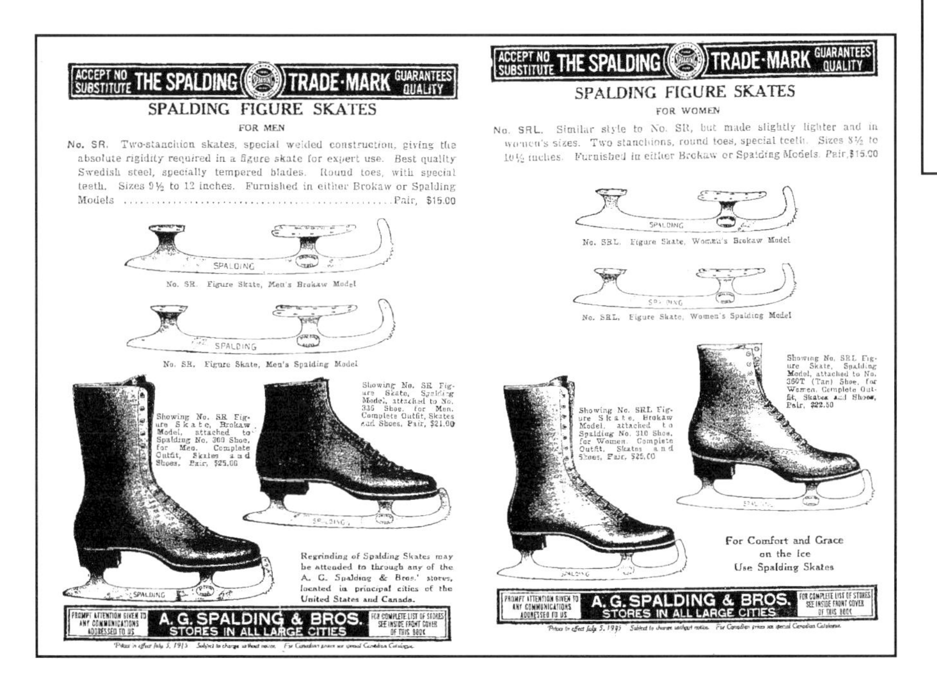

Figure 85. Advertising sheets illustrating the Spalding Company's figure skates for men and women.

The following pages illustrate several American handmade and factory made skates. Many have exceptionally large curled blades and others are truly one-of-a-kind hand crafted, graceful looking skates. Some illustrate the exotic wooden platforms such as tiger strip maple, bird's eye maple, walnut and rosewood. All of them have brief descriptions outlining their features. The photographs of the skates were taken in the four seasons, winter, spring, summer, and fall.

Figure 87. An outstanding, classy pair of handmade skates. The runners are very decoratively cast in bronze with beautiful turn-up curls and knob finials. The three arched stanchions are cast integrally with the runner. Note the elevated bronze piece supporting the arch of the foot. The wooden platform is arched up and attached to the stanchion. The plan view of the platform is carved in the shape of an hourglass. There are three cast pewter escutcheons inlaid flush with the curved edge of the platform and screwed on with handmade screws. These three slots accommodate the leather straps for fastening. Long screws are mounted in the heel to fasten the boot securely to the skates. This highly skilled craftsman was very knowledgeable in pattern making, as well as bronze and pewter casting procedures. They are a super pair of skates. Length 11 3/4", Width 2 1/2", Platform height 2 1/8", Curl diameter 2 1/2", c. 1850. Class H: $1000-1500.

Figure 88. This is a one-of-a-kind handmade pair of skates that is illustrated on the front cover. The decoratively engraved runners were made by a very skilled blacksmith. They were hand forged and twisted several revolutions while they were red hot. He also forged square pyramid-shaped finials at the tip to set them off. He decorated the runner blades by engraving and hand punching them. He used a variety of tools to get a design of diamonds, stars, and a border. The platforms are made from walnut with an adjustable handmade spike in the heel to help support the boot. There are three nail barbs in the toe to prevent slipping. The hand sewn leatherwork on the harness cups is of high quality. They truly are a work of art. The skates are accompanied by a brass, cobalt blue skater's lantern. Length 13", Width 2 1/8", Platform height 1 3/4", Curl height 4 1/2 ", c. 1840. Class H: $1000-1500.

Figure 89. A unique pair of skates without stanchion supports, like a springboard. The blacksmith's work on the runners is very good. They look like sled runners. The ironwork on the runners has the look and quality of the French style. The platforms are screwed to the upper runner bar underneath. Original leather cups are screwed to the platform and leather straps of rawhide are used for fastenings. The skates are near a brass, green globed skater's lantern ready to go. Length 12", Width 2 1/4", Platform height 2", c. 1840. Class F: $500-750.

Figure 90. Handmade skates with nice sized curls on the fronts, which have brass acorn finials at the tips. The platform has three straps for fastening the boot and metal spikes in the heel for security of boot slippage. A red skater's lantern is sitting beside the skates, both waiting to take a spin on the pond. Length 12", Width 2 1/4", Platform height 1 1/2", Curl diameter 4 3/8", c. 1850. Class E: $350-500.

Figure 91. Handmade skates made with forged runners and brass acorn finials. The platforms are very thick and sawed out in the shape of the foot. The platform is fastened to the runner with a riveted pin. There are two leather straps to fasten the skates to the boots. A dark green brass skater's lantern is sitting next to the skates. Length 11 3/4", Width 2 1/4", Platform height 1 7/8", Curl diameter 3 3/4", c. 1850. Class D: $200-350.

Figure 92. A beautifully proportioned handmade pair of skates with extra large forged curls and finials. The platform is tapered down to the top of the runner in front. It has screws at the heel to prevent boot slippage. There are three leather straps for fastening. Length 15 1/2", Width 2 1/2", Platform height 1 7/8", Curl diameter 5 5/8", c. 1840. Class F: $500-750.

Figure 93. A nice factory made pair of women's skates. The runners were made by forging a thick piece of steel into a "U" shape form simulating the gutter or hollow ground blade. The runners are bent up into a curl form on the front. The two stanchions are fabricated out of brass and screwed to the wooden platform. The leather fastenings cover the entire toe area on the front and there are leather heel cups on the rear. The leathers are screwed to the edges of the platform. Length 11 1/4", Width 3", Platform height 1 3/4", Curl diameter 4 1/4", c. 1860. Class D: $200-350.

Figure 94. A smaller pair of skates, probably women's. The forged runner has a small curl and brass acorn finial. The wooden platform is made of cherry wood. The platform is hourglass-shaped and has three leather straps for fastening. The platform heel has a metal spike for support. The heel leather harness is hand stitched with linen thread. Length 12 1/4", Width 2", Platform height 1 5/8", Curl diameter 2 7/8", c. 1860. Class D: $200-350.

Figure 95. A nice looking pair of hand forged metal skates. The runners are wrought with decently sized curls on the front. The heart-shaped foot plate and the heel plate are forged to the stanchion runner. The foot plates are made to have leather straps attached with rivets, as evidenced by the rivet holes and eight remaining rivets. A large screw is forged into the heel for boot support. A brass, cobalt blue skater's lantern accompanies the skates. Length 12 1/4", Width 2 3/8", Platform height 1 1/2", Curl height 2 7/8", c. 1850. Class D: $200-350.

Figure 96. Handmade skates with super big curls. The runners have a pair of eye catching "eyes" forged at the tips as finials. The walnut platform has two leather fastenings and an adjustable square pointed spike in the heel for securing the boot. Sharp barbs on the foot platform prevent slippage. This is a "keeper" if you are fishing. Length 3 1/4", Width 2 1/8", Platform height 1 7/8", Curl diameter 6 1/4", c. 1850. Class F: $500-750.

Figure 97. An outstanding pair of hand crafted skates with beautiful scrolled curls. The hand forged runner has a hole drilled in the top front used to attach it to the platform with a metal staple. The staple is threaded up through the platform and clinched over on top to secure the runner to the platform. The platform has two straps for fastening, as well as spikes in the heel area to secure the boot. The runner is exceptionally thick, 3/8", and heavy. Length 13 1/2", Width 3", Platform height 1 7/8", Curl diameter 3 5/8", c. 1850. Class G: $750-1000.

Figure 98. Skates with large, unusual curls on the fronts. The runner blade is rectangular where it meets the ice and is then forged round towards the tip. The curls give the appearance they are flapping in the wind from skating so fast! The platforms are carved like one's foot in the shape of the arch. There are two leather fastenings and a heel spike to secure the boot. They appear to be an early pair of factory made skates. Length 13 1/4", Width 3 1/8", Platform height 2", Curl diameter 5 1/2", c. 1870. Class E: $350-500.

Figure 99. A quality pair of factory made, low profile racing skates. The runners are perfectly flat for fast forward skating. The platform has three leather straps for fastening to the boots. There is a front toe plate on top, a plate mounted with sharp spikes in the center, and a decorative heel screw supporting the boot at the rear. Length 15 1/2", Width 2 1/4", Platform height 1 5/8", c. 1870. Class B: $25-100.

Figure 100. This pair of iron skates is an example of good blacksmithing work. The hand forged curls are very bold looking and set forward from the platform. Three rectangular stanchions are inlet into the metal platform and welded in place at the blacksmith's forge. Copper rivets hold the leather straps to the platforms. Length 14 3/4", Width 3 1/4", Platform height 1 1/4", Curl diameter 4 5/8", c. 1850. Class F: $500-750.

Figure 101. Very heavy hand wrought metal skates. They are unusual because the runners were forged out of old worn out rasps. The rasps were heated, hammered smooth, and then forged into the shape of a skate runner curl. Some rasp marks can still be seen on the surface. The stanchions were forge welded to the runner and plates. The foot plates are very heavy and painted a dark green. Spikes are mounted in the heels to support the boot. There are notches cut into the front platform for the straps to go through for fastening. A person would quickly tire from constant lifting of these heavy skates. Length 12", Width 2", Platform height 1 3/4", Curl diameter 4", c. 1870. Class D: $200-350.

Figure 102. A small pair of handmade skates that has a very distinctive prow or curl. The hand wrought curl has been forged into the shape of a "Viking" ship. The maker probably had roots in one of the Scandinavian countries. The platform has traces of old red paint remaining and has a spike in the heel for boot support. Leather straps on the toe area and leather thongs on the rear were used for fastening. Length 11 1/4", Width 2 1/4", Platform height 1 1/2", Curl height 2 3/8", c. 1860. Class E: $350-500.

Figure 103. A primitive, handmade pair of skates with a bold sweep on the front of the runner, ending in dainty scrolled, turn over curls. The platform is anchored to the runner by a pin driven through it. There are two leather straps for fastening the boots and metal nails for securing the foot. The oak platform is very thick and carved by hand. The photo was taken inside a log cabin with candlelight. Length 13 1/2", Width 2 1/2", Platform height 1 7/8", Curl diameter 2 3/8", c. 1860. Class D: $200-350.

Figure 104. A pair of large figure skates with high runner blades and a curved profile on the bottom. The cast steel runner has a radius on front and back so the skater can skate forward and backward. The platform is made from maple and has two leather straps and a screw at the heel for fastening. They are a common "plain Jane" pair of skates. Length 13 3/8", Width 2 3/8", Platform height 1 3/4", c. 1890. Class B: $25-100.

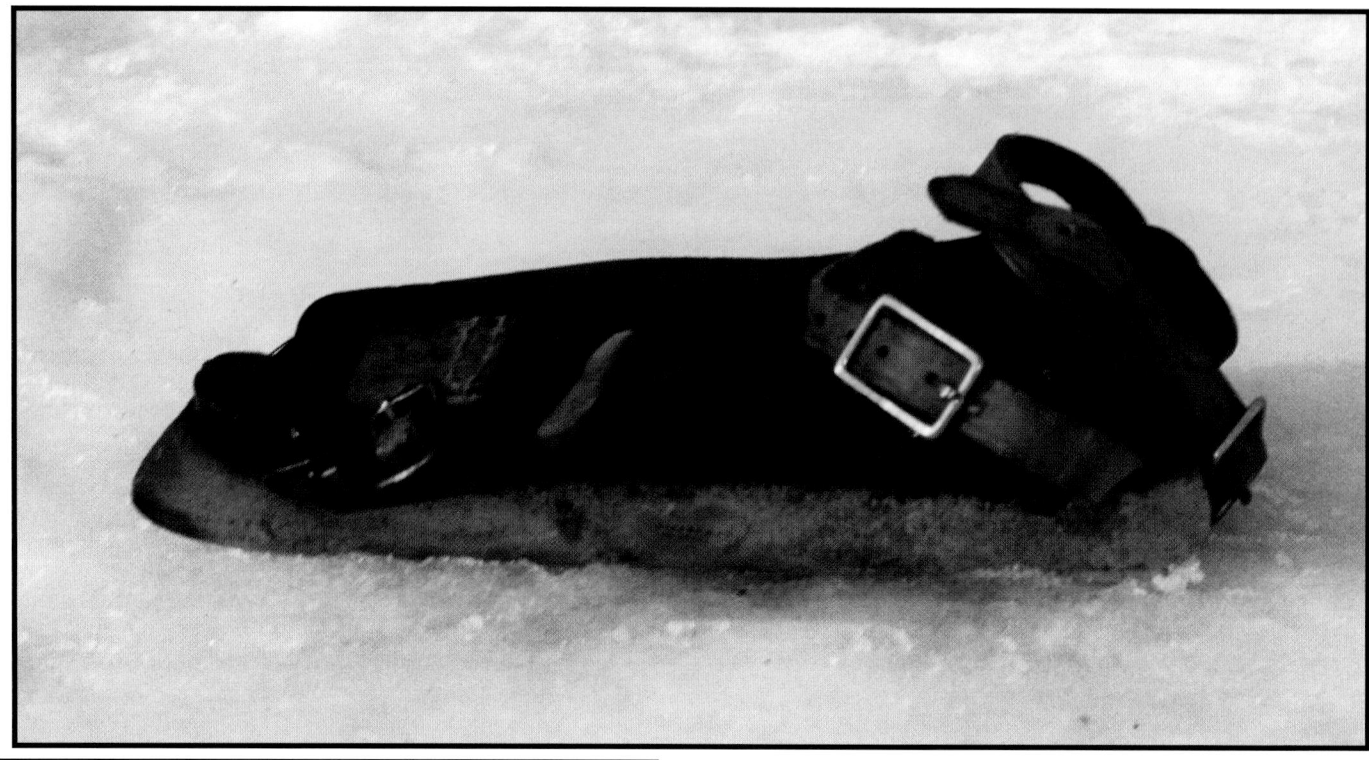

Figure 105. Factory made pair of figure skates with a radius on both front and rear of the runner. The wooden platform is made from rosewood and is contour carved to fit the arch of the foot. The platform has screws in the heel and barbed spikes in the foot plate to prevent slippage. Two leather straps are used for fastenings. Length 11 1/2", Width 2 1/4", Platform height 1 3/4", c. 1880. Class D: $200-350.

Figure 106. Skates with heavy cast steel runners, 5/16" thick. The platform has an hourglass shape extending right at the curl for extra support. Two leather fastenings and a spike in the heel were used for support. Length 11 3/4", Width 2 1/2", Platform height 2", Curl diameter 4 1/2", c. 1850. Class D: $200-350.

Figure 107. A pair of small, hand forged metal skates with curls on the runners. The runner blades were forged out of two old files with some of the file teeth still showing. The iron platform is supported by two metal stanchions forge welded in place. The platform is very narrow with metal spikes mounted in the heel. The leather straps are attached to the platform by copper rivets. Length 11 1/2", Width 1 5/8", Platform height 1 1/2", Curl diameter 3 3/4", c. 1850. Class D: $200-350.

Figure 108. A pair of large hand forged skates with nice big curls. The runner blades were forged from old rasps, the tang forming the curl. The wooden platform is tapered in the front and extends and joins the runner at the curl. The platforms are painted red over an old white undercoating. From the wear and tear of using the skates both the red and white paint is showing. There are three leather fastenings, a heel spike, and barbs on the foot to secure the skate to the boot. These skates are much more attractive in the flesh than in the photo. Length 13 7/8", Width 2 1/2", Platform height 1 5/8", Curl diameter 4 3/8", c. 1850. Class E: $350-500.

Figure 110. Handmade skates with forged runners and small curls. The blacksmith's forge welding and decorative work on the three runner stanchions is quite good. The wooden platform is painted a dark green with some wear in the proper places. Three leather straps were used for fastenings. A red globe skater's lantern is shown next to the Christmas green skates. Length 11 5/8", Width 2 1/4", Platform height 2", Curl diameter 2 1/4", c. 1850. Class D: $200-350.

Figure 109. Early cast steel runners with respectful curls on the fronts. The platform is tapered at the front and joins the curl for extra support. The platform is shaped like an hourglass. The original harness cups are attached at the rear and to the front. The heel has a spike and the foot has sharp barbs to secure the boot. Length 12 1/4", Width 2 1/4", Platform height 1 3/4", Curl diameter 4 3/8", c. 1850. Class D: $200-350.

Figure 111. Factory made skates manufactured by the Douglas Rogers and Co., Norwich, Connecticut, and patented on March 17, 1863. This pair has an unusual and attractive curl on the front that attaches to the platform front. The stanchions are riveted to the runners and screwed to the underside of the platform. The platform has three leather straps for fastening and a metal spike in the heel for support. Length 11", Width 2 1/4", Platform height 1 7/8", Curl height 2 3/8", c. 1863. Class D: $200-350.

Figure 112. Small handmade skates with a small turn-up prows. The runners are hand forged and hollow ground on the bottom. The platforms are hand carved and have heel screws to prevent slippage of the boot. Two leather straps were used for the fastenings. A brass skater's lantern accompanies them for an evening spin on the pond. Length 11 3/4", Width 2 1/4", Platform height 1 1/2", Curl height 2 1/4", c. 1840. Class C: $100-200.

Figure 113. Women's factory made skates with nice forged runners and closed curls on fronts. The runners have two brass bell-shaped stanchions attaching them to the platform. Someone improvised the front platform with metal support clips to hold the toe onto the platform. The rear has leather cups with reinforced metal backings screwed to the platform edge. Old yellow paint is still evident. Length 10 3/4", Width 2 3/8", Platform height 1 5/8", Curl diameter 1 7/8", c. 1870. Class C: $100-200.

Figure 114. An early pair of wooden platform speed skates retrofitted by marrying them with a pair of Union Hardware metal skates, attached to the rear! Note the cloverleaf stamp and adjustable clips on the toe. The fronts of the wood platforms were sawed off and then replaced by the front of the metal skates. The metal add-ons were attached at the rear with copper rivets and the fronts were attached to the curls with brass supports. The skates are put together pretty well by the innovator. He thought he could help solve the problem of his boots coming off by using the newer metal clip design on his old wooden skates. A cobalt blue skater's lantern is sitting next to the skates. Length 13 1/2", Width 3", Platform height 1 3/4", c. 1890. Class D: $200-350.

Figure 115. Primitive handmade skates. The overall appearance of the front runner and platform resembles a steamship's bow. The oak platform is hand carved with two leather straps for fastenings. The platform extends beyond the runner at the rear and has long screws for heel support. Length 13", Width 2", Platform height 1 5/8", c. 1870. Class B: $25-100.

Figure 116. A beautiful pair of medium size figure skates with nice big curls and brass acorn finials at the tips. The platform has three sets of straps for fastening to the boot. Adjustable metal screws in the heel and three nails on the foot pad prevent boot slippage. Length 12 5/8", Width 2 5/8", Platform height 2", Curl diameter 5 5/8", c. 1850. Class F: $500-750.

Figure 117. Classy skates with a curl on both the front and rear of the runner. Rectangular stanchions are forge welded to the runners. The platform is carved out of a beautiful piece of bird's eye curly maple. The platform has metal spikes at the toe plate and long threaded screws at the heel for securing the boot. The original leather fastenings still have linen thread stitching on them. Length 12", Width 2 5/8", Platform height 2", c. 1850. Class F: $500-750.

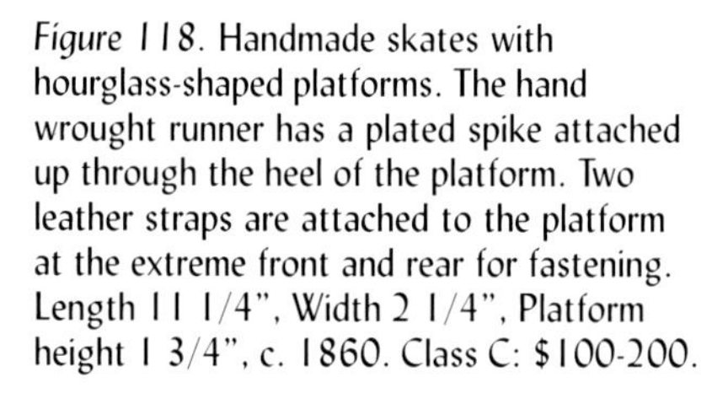

Figure 118. Handmade skates with hourglass-shaped platforms. The hand wrought runner has a plated spike attached up through the heel of the platform. Two leather straps are attached to the platform at the extreme front and rear for fastening. Length 11 1/4", Width 2 1/4", Platform height 1 3/4", c. 1860. Class C: $100-200.

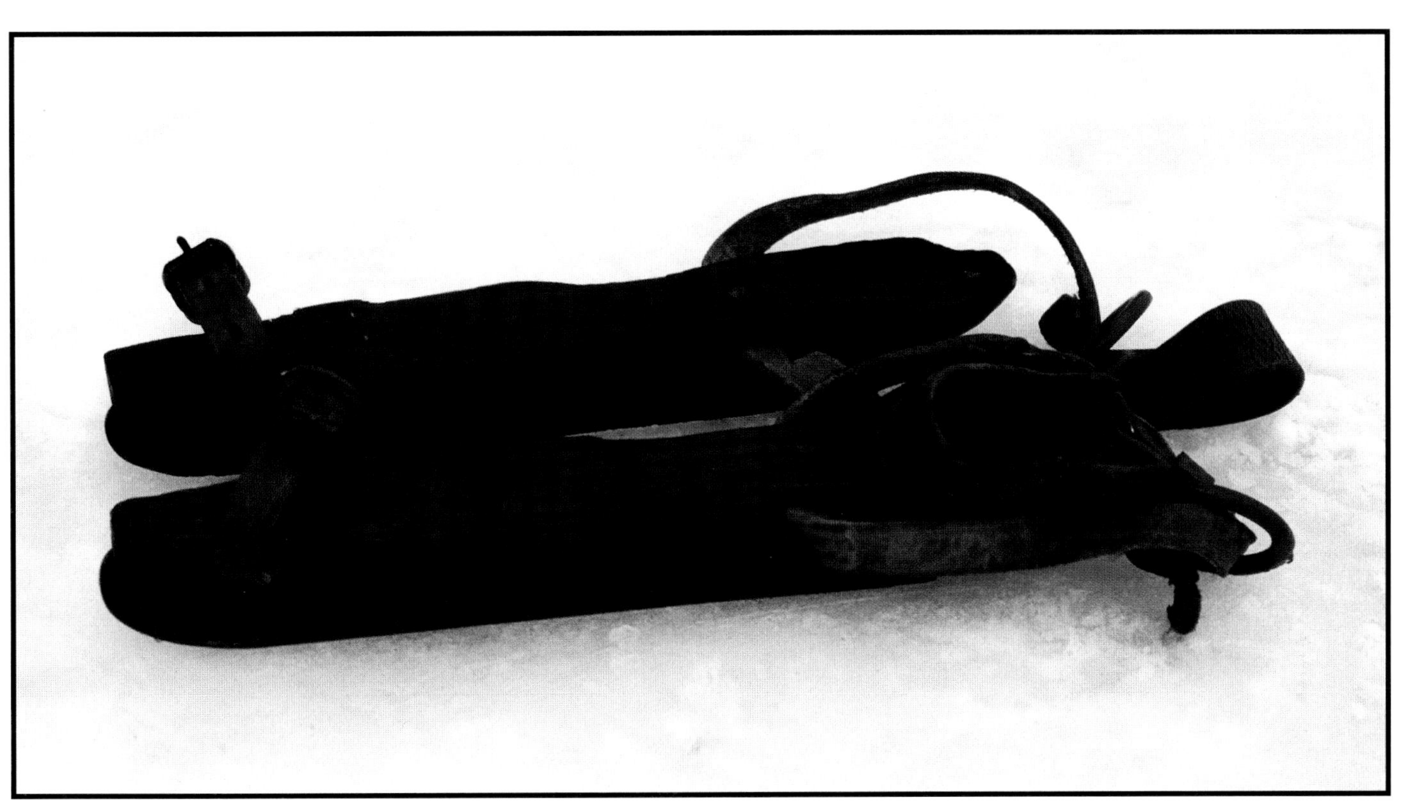

Figure 119. An ordinary pair of ice skates with metal barbs on the toes and spikes on the heels. The rear has a harness strap and a single strap on the toe. The platforms are hand carved and made of oak. The Queen of England would not have owned this pair of skates. Length 11 1/4", Width 2 1/2", Platform height 1 5/8", c. 1870. Class B: $25-100.

Figure 120. A very attractive pair of handmade skates, with long hand forged runners ending in dramatic and forward-looking, double scrolled curls. The platform has an hourglass shape with two leather loops in front and harness fastenings at the rear. The runners are fastened to the platforms with metal pins at the front. There is still evidence of old black paint on the skirt of the platforms. Length 13", Width 2 1/8", Platform height 1 1/4", Curl diameter 2", c. 1840. Class E: $350-500.

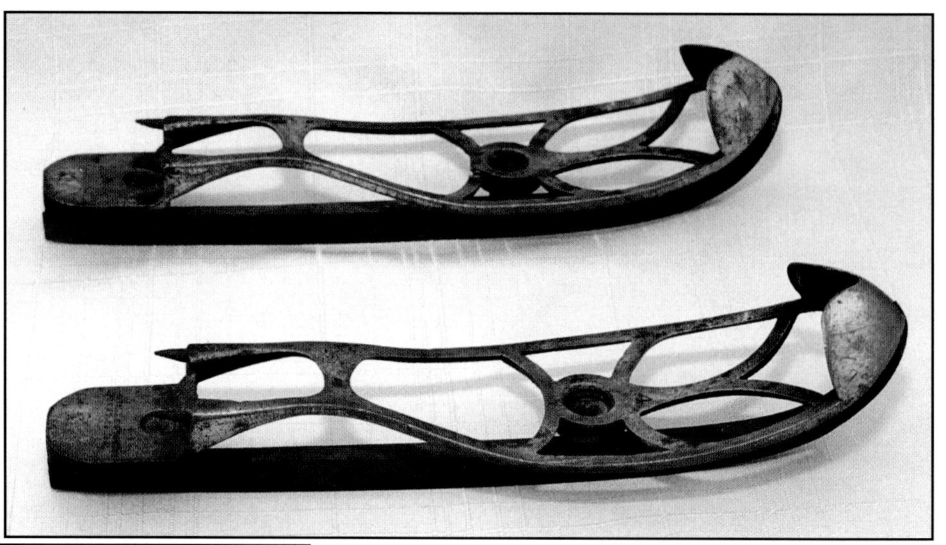

Figure 121. Decorative looking cast brass skates. They are factory made and stamped with the Smirley's Patent, dated June 25, 1861, Number 4. Note the toe supports on the front and the horizontal spikes on the rear. The cast brass skates were originally nickel plated, but much of it has since worn off. Length 10 3/4", Width 2 3/8", Platform height 1 3/8", c. 1861. Class D: $200-350.

Figure 122. Three outstanding pairs of curled skates accompanied by three skater's lanterns. They are waiting for the sun to go down to take a spin on the pond.

Figure 123. A wonderful pair of handmade skates that clearly show the runners were hand forged out of old rasps. The other side of the runners have the file marks on them from the combination rasp/files they were made from. The blacksmith forged a nice curl from the tang of the file! The platforms were hand carved and morticed at the bottom for the two leather strap fastenings. The platforms are painted with an old red paint, the majority of which still remains. This is a very desirable pair of skates. Length 14 1/4", Width 2 7/8", Platform height 1 7/8", Curl diameter 3 3/4", c. 1850. Class F: $500-750.

Figure 124. An outstanding pair of handmade skates with tiger stripe maple platforms. The platforms were missing and have been replaced. The hand forged runners have nice curls and turned over finials. The runner is attached to the platform by two brass washer clips riveted to the stanchions and then screwed to the platform. A brass skater's lantern with a cobalt blue globe is sitting next to them. Length 12", Width 2", Platform height 1 1/2", Curl diameter 5 3/8", c. 1850. Class F: $500-750.

Figure 125. Handmade figure skates having exceptionally large curls with brass finial at the tip. They are probably women's skates since the platforms are rather short. The hand carved platforms have screws in the heel to secure the boot. Two leather straps were used for fastenings. Length 11 3/4", Width 2 1/8", Platform height 1 5/8", Curl diameter 5 1/2", c. 1850. Class F: $500-750.

Figure 126. An unusual pair of metal ankle supported skates. They were manufactured by the Barney and Berry Skate Co., Springfield, Massachusetts., and patented May 2, 1876, Oct. 18, 1881, Feb. 12, 1884, March 1, 1887, and Aug. 11, 1896. They have hinged and adjustable height support straps to help support the skater's ankles. They also have adjustable heel and toe clips to tighten onto the boots. A differently designed cone top skater's lantern is sitting next to the skates. Length 13", Width 3 1/4", Platform height 1 3/4", c. 1880. Class D: $200-350.

Figure 127. Decorative cast iron skates. The front foot pad has cast slots for the leather straps to go through. The rear heel pads are heart-shaped and have a cast knob in the center of them. This knob fits into a female clip screwed to the bottom of the boot for attachment. Length 9 3/4", Width 2 3/4", Platform height 1 3/4", c. 1880. Class C: $100-200.

Figure 128. This pair of figure skates appears to be factory made. The runner has a nice graceful curl with a brass finial at the tip. Ninety percent of the old orange paint still remains on the platform. The skates have a square-shaped spike in the heel to secure the boot heel and three leather straps for fastening. Length 11 3/4", Width 2", Platform height 1 3/8", Curl diameter 4 3/16", c. 1860. Class E: $350-500.

Left: *Figure 129*. A very early pair of cast iron skates. The bottom of the runners are cast thick, 3/8". The curl on the front makes a complete sweep back. The top of the runner has two round washer type clips to fasten to the platform. The platform extends forward and meets the runner for extra support. The platform has a spike at the heel for securing the boot and two leather straps for fastening. Length 13", Width 2 5/8", Platform height 1 7/8", Curl diameter 4 1/2", c. 1860. Class D: $200-350.

Right: *Figure 130*. Handmade figure skates with a nice wrought curl on the front. The forge marks are evident on the curl. The wooden platforms are unusually thick and have three slots for leather fastenings. The platform has sharp spikes at the toe and a long screw at the heel for securing the boot. Length 11 3/4", Width 2 3/8", Platform height 1 5/8", Curl diameter 3", c. 1840. Class D: $200-350.

Left: *Figure 131*. Handmade skates with a very high runner blade. They have nice curls and finials wrought at the tips. The runner on this figure skate is hollow ground. The platform has two spikes and a hand filed screw at the heel for securing the boot. A leather harness at the rear and loops at the front were used for fastening. The skate has a Dutch look to it. Length 13", Width 2 1/8", Platform height 1 5/8", Curl diameter 3 3/4", c. 1850. Class D: $200-350.

Right: *Figure 132*. A handmade pair of skates with a curved runner on the bottom. The runner has a groove forged on either edge as well as being hollow ground on the bottom. The runner has a beautiful curl on the front with a forged finial at the tip. The platform has three slots for leather fastenings and a spike on the foot area to secure the boot. Length 13 7/8", Width 2", Platform height 1 1/2", c. 1840. Class E: $350-500.

Figure 133. Long racing skates with turn-up curls on the front. They were manufactured by the J.L. Whelpley Skate Co., Boston, Massachusetts. The maple platform has four slots for leather fastenings and is attached to the runner with two screws. The top of the platform has two sharp spikes and a screw in the heel to secure the boot. Length 16 1/8", Width 2", Platform height 1 1/2", Curl height 1 7/8", c. 1860. Class D: $200-350.

Figure 134. An unusual pair of spring skates with no stanchions. The runner is fastened in the front and rear of the platform with large brass screws. The platform has three decorative plates mounted on it with sharp spikes and screws to support the boot. The platform is contour-carved to fit the foot arch. The runner has a pronounced round bottom for figure skating. Length 12 3/8", Width 2 3/8", Platform height 2", Curl height 2 1/8", c. 1840. Class E: $350-500.

Figure 135. A decorative pair of factory made skates. The platform is cast iron and the runner is forged steel with a curl and knob at the tip. The runners are fastened to the platform with cast stanchions. The heel has a sharp pin to support the boot. The fastenings consist of a harness with rings at the rear and straps at the toe. Eighty percent of the original black paint still remains. Length 10 1/4", Width 2 1/4", Platform height 1 1/2", Curl height 2 1/4", c. 1890. Class C: $100-200.

Figure 136. Handmade racing skates. The forged runner has a nice turn up curl that resembles a Viking ship. The platform is made of cherry wood and has two leather straps for fastening. The heel has a sharp spike to secure the boot. Length 13 7/8", Width 1 7/8", Platform height 1 1/4", Curl height 2 1/4", c. 1840. Class D: $200-350.

Figure 137. Figure skates with curved runners on the bottom. They were manufactured by the C.H. Pratt Co., Roxbury, Connecticut. The platforms are made of rosewood with decorative brass plates inlaid in them. The platform front has two sharp nails and a pin in the heel to secure it to the boot. There are three leather straps used for fastening. Length 11 1/8", Width 2 1/2", Platform height 1 7/8", c. 1860. Class D: $200-350.

Figure 138. An unusual pair of factory made skates with leather shoe tops. The shoe is attached to the wooden platform and metal runner. The shoe tops have a red leather border sewn around them, as well as down the lace skirts. It appears to be a women's skate. The runner has a big diameter hollow or gutter cut on the bottom. Length 10 3/4", Width 2 1/4", Platform height 1 3/8", Total height 7 1/2", c. 1880. Class C: $100-200.

Figure 139. Factory made figure skates with a small closed curl at the front. The runners were forged to the three stanchions, which were attached to the platforms with brass key nuts. The three stanchions are cone-shaped and eight sided. The platforms have screws at the heels and nails at the foot plate to support the boot. Leather straps were used at the toe area. The skates were manufactured by the H. Clark's Skate Co., Jordon, New York. The illegible patent stamps on the wood appear to be March 18, 1861, and April 23, 1862. Length 12 1/8", Width 2 3/4", Platform height 1 3/4", Curl diameter 2", c. 1860. Class D: $200-350.

Figure 140. This early pair of handmade skates are women's or children's skates judging from their size and appearance. The runner is forged into a unique prow, which resembles a Scandinavian ship. The oak platform is hand carved, as the chisel marks are still apparent. The runner is attached to the platform with a clamp on the front and heel screw on the rear. A pointed handmade adjustable screw is on the heel and nails are on the front to support the boot. Two leather straps were used for fastening. Length 10 3/8", Width 1 7/8", Platform height 1 1/2", Curl height 1 3/4", c. 1850. Class D: $200-350.

Figure 141. Factory made long racing skates with holes punched in the runners. The sixteen holes were punched in the blades to lighten them for skating, not to dissipate the heat from the runners of a fast skater! The platform has a support plate at the nose, foot area, and heel. There are two spikes at the foot area and a screw at the heel to support the boot. They may have been made by the Whelpley Co., Boston, Massachusetts. Length 18 1/4", Width 2 1/4", Platform height 1 5/8", c. 1870. Class D: $200-350.

Figure 142. A pair of medium size skates, probably women's. They have a small curl on the front and round knob finial. The runner has two stanchions with a spike coming up through the platform at the heel. The platform has two leather straps for fastening. Length 11 1/2", Width 2 1/4", Platform height 1 5/8", Curl diameter 2 1/4", c. 1860. Class C: $100-200.

Figure 143. Factory made women's skates with enclosed leather toes and heels. The wooden platforms are very narrow at the arch. The leathers are held onto the platform with brass plates and screws. The straps are hand sewn with linen thread. The runners have a nice small curl up attached to the platform top. There are two brass stanchions attaching the runners to the platform. Length 11", Width 2 3/4", Platform height 1 3/4", Curl diameter 1 5/8", c. 1860. Class D: $200-350.

Figure 144. Later boot top skates made by the Union Hardware Co., Torrington, Connecticut. The stanchions are screwed to the boot soles, which solved the age old problem of skates coming loose and off the foot. Length 11 1/2", Width 3", Platform height 2 1/8", Total height 9 1/4", c. 1905. Class B: $25-100.

Figure 145. Boot top skates manufactured by the Barney and Berry Co., Springfield, Massachusetts. They have four stanchions on the runner, which are permanently riveted to the boots. The name "Filene's" is stamped on the bottom of the leather soles. Length 12 1/4", Width 3 3/8", Platform height 2", Total height 10 1/2", c. 1905. Class B: $25-100.

Figure 146. Handmade skates with beautiful curls on the runners resembling an old Viking ship. The platform is in the shape of an hourglass and still has traces of old red paint. Leather loops and thongs are used for fastenings. A spike in the heel is used for securing the boot. Length 11 3/8", Width 2 1/4", Platform height 1 3/8", Curl height 2 3/8", c. 1820. Class E: $350-500.

Figure 147. Medium sized skates with small turn over curls that join the platforms at the fronts. The runners have two stanchions attaching them to the platforms. The platforms have three leather straps for fastening and a screw in the heel for supporting the boot. They were manufactured by the Douglas Rogers and Co., of Norwich, Connecticut. They are stamped "steel tempered patented March 17, 1863." Length 11 1/4", Width 2 1/2", Platform height 1 3/4", Curl height 2", c. 1863. Class C: $100-200.

Figure 149. Factory made skates made by the Union Hardware Co., Torrington, Connecticut. The runner has a prow and decorative stanchions. Brass heel cups are screwed to the platform skirt. The platform has a spike on the heel and a wide strap on the toe for boot fastening. Length 10 5/8", Width 2 1/4", Platform height 2 1/4", Curl height 2 1/2", c. 1880. Class C: $100-200.

Figure 150. This handmade pair of figure skates, with curved bottom runners, was forged by the blacksmith from old worn out files. One can still see the file teeth on the blades. The platform has two sharp nails on the toe and a spike on the heel to hold the boot. Two leather straps were used for fastening. Length 11 1/8", Width 2 1/4", Platform height 1 3/4", c. 1860. Class D: $200-350.

Figure 148. Factory made skates by H. Clark's Co., Jordon, New York. The skate is stamped "patent March 1, 1861, size 9 1/2." The platform has a raised toe area and a screw in the heel to support the boot. The runners are fastened to the platform by two brass bell-shaped stanchions. Two leather straps were used for the fastenings. Length 10 1/8", Width 2 1/2", Platform height 1 1/2", Curl height 2", c. 1860. Class D: $200-350.

Figure 151. A handmade pair of skates with a round toe runner forged from an old, worn out file. The teeth can still be seen on the blades. The platform has two sharp nails on the toe area and a screw at the heel to support the boot. Two leather straps were used for fastening. Old red paint is still seen on the wooden platform. Length 11 3/4", Width 2 5/8", Platform height 2", c. 1860. Class D: $200-350.

Figure 152. Factory made metal skates with two platforms for the foot and heel. The runner has a curl with a knob finial, and is attached to the platform by riveting two round stanchions between them. There is a small star on the rivet heads. There are three spikes on the foot platform and heel clips to hold the boot in place. Leather straps riveted to the platforms serve as the fastenings. Length 11 1/2", Width 2 1/2", Platform height 1 7/8", Curl diameter 2 3/4", c. 1860. Class D: $200-350.

Figure 153. Old handmade skates with a nice forged curl on the front. The blacksmith forged the runner out of an old worn file, forming the tang into the curl. The wooden platform is very wide and held to the runner by a split rivet clinched over. The platforms have upper leathers nailed and screwed onto the skirt forming the boot. The leather straps and buckles were hand stitched together with linen thread. Length 13 3/8", Width 3 3/4", Platform height 1 5/8", c. 1830. Class D: $200-350.

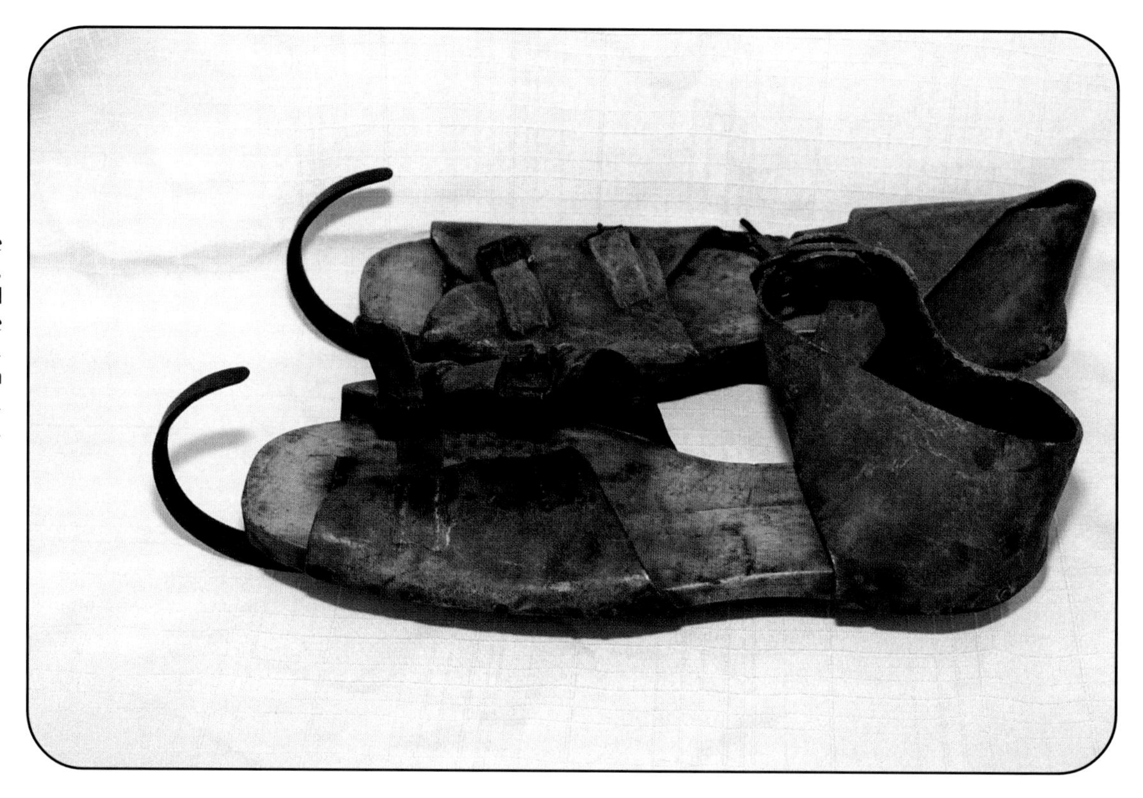

Figure 154. These skates have unusual forged runners with large clamps embedded into the platform fronts. The runner has a curl with a uniquely shaped finial made of brass. The platforms are rather thick and have a step down for the heel. Metal pins in the center of the heel secure the boot heel. Two leather straps are used for fastening. Length 12 1/4", Width 2 1/4", Platform height 1 1/2", Curl height 2 1/2", c. 1860. Class D: $200-350.

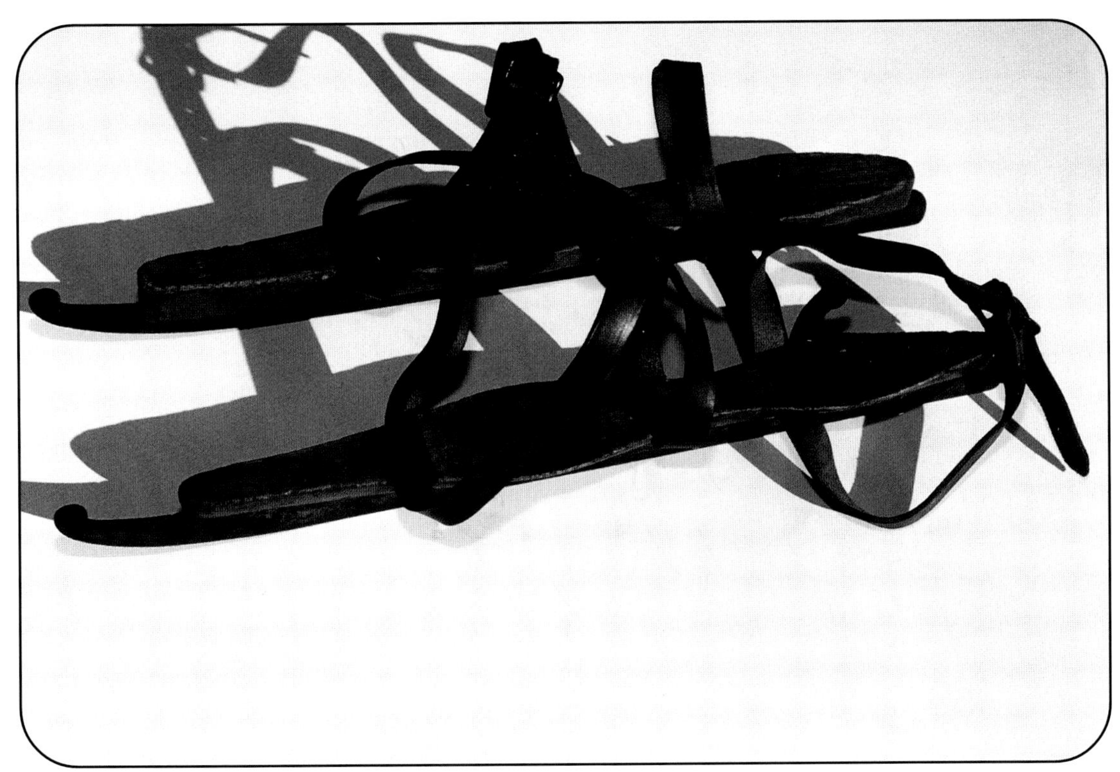

Figure 155. Very long speed skates with nice finial knobs on the fronts. The skates are very narrow with sharp nails on the toe area and spikes at the heels. The platforms have three morticed slots for leather strap fastenings. There are long screws at the heel for boot support. Length 16 1/4", Width 1 3/4", Platform height 1 3/8", c. 1870. Class C: $100-200.

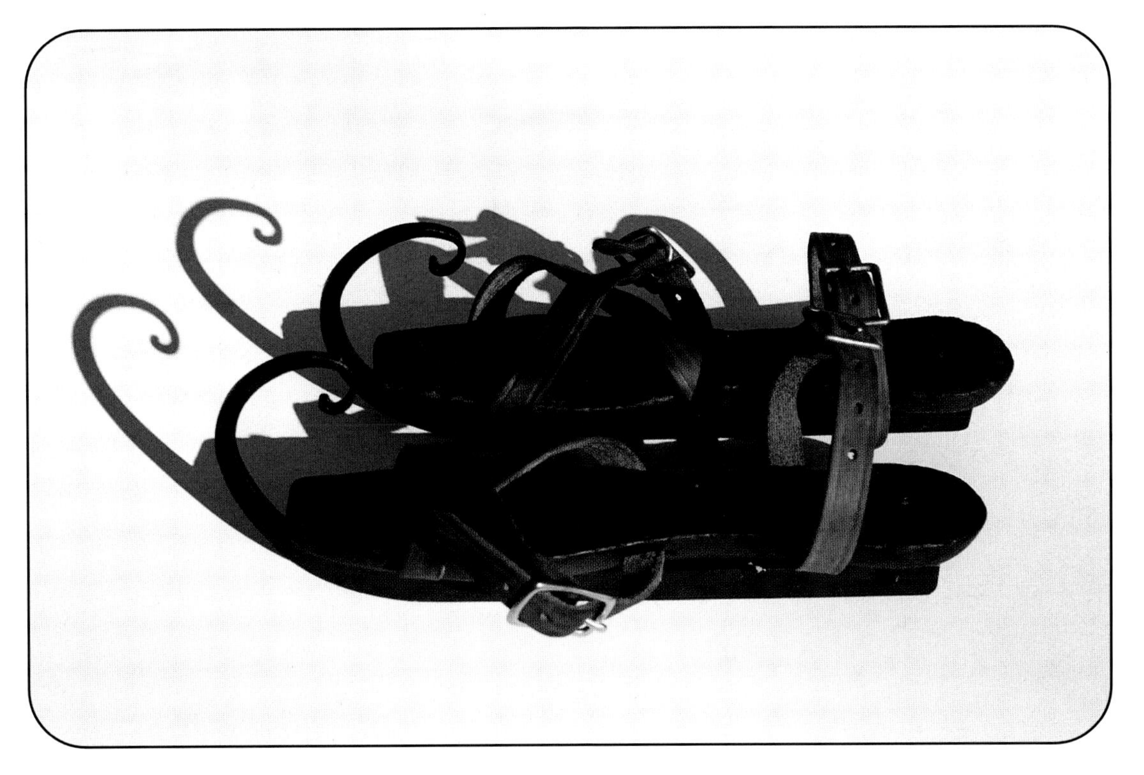

Figure 156. Handmade skates with nice forged curls at the tops. The decorative curls are embellished in the sun's shadow. The runner has two round turned stanchions attached to the platform. The walnut platform has three leather slots for fastenings to the boot and heel screws to prevent slippage. Length 11 1/8", Width 2 3/8", Platform height 1 5/8", Curl diameter 4 1/4", c. 1860. Class E: $350-500.

Figure 157. Handmade skates with forged curls. The tapered platform extends entirely up to the curl for more support. The platform has two slots for leather fastenings. A harness strap arrangement is riveted in the rear and straps are in the front. The platform has an inlet plate holding a spike to support the heel. Length 11 1/4", Width 2 3/8", Platform height 1 5/8", Curl diameter 4 1/2", c. 1850. Class D: $200-350.

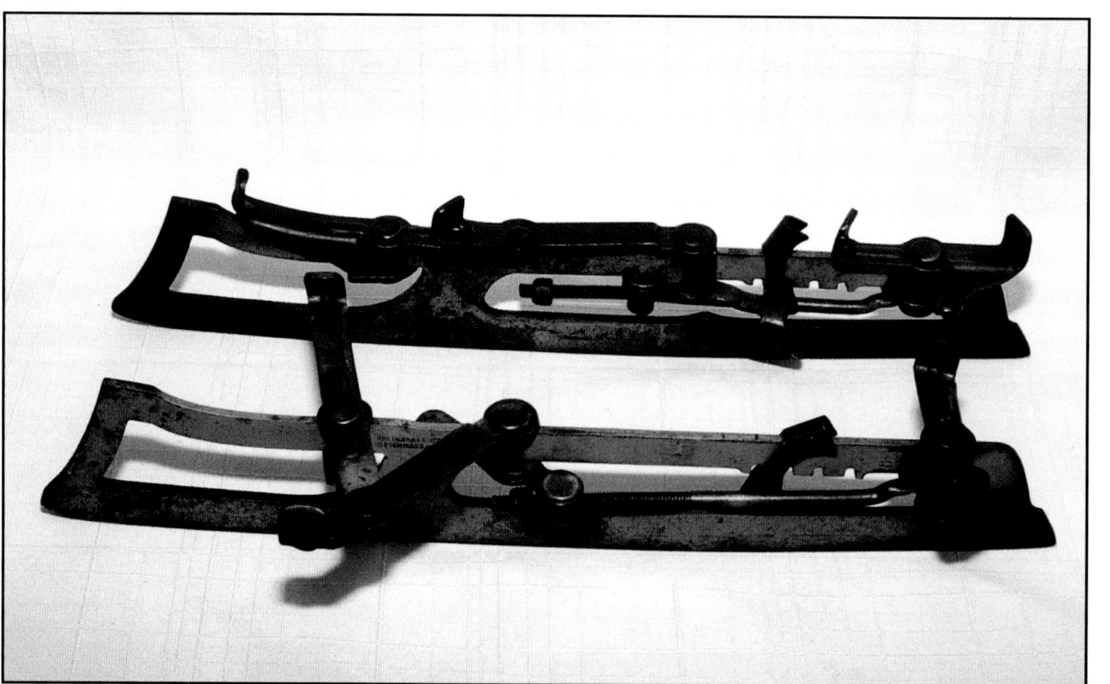

Figure 158. Unusual metal skates manufactured by the Artin Skate Co., Boston, Massachusetts, and patented Nov. 7,1905. The heel clips are adjustable for different lengths of boots. The unique feature of these skates is that they fold completely up into one plane, as shown in the photograph, for easy transport. There is a cam action lever that tightens the clips to the toe as they are opened up for use. Length 13", Width 5/8", Platform height 2", c. 1905. Class C: $100-200.

Figure 159. Small factory made metal skates with no support platforms or stanchions. This spring skate design was made by the J.A. Lang Co. The oval foot plate has two sharp barbs to hold the boot from sliding and long heel screws to secure the heel. Leather straps were used on the foot area for fastening. Length 10", Width 2 3/4", Platform height 1 1/2", Curl diameter 2", c. 1870. Class C: $100-200.

Figure 160. Women's factory made skates. They are cast steel and have decorative foot plates with metal barbs to secure the toe area. A male knob on the heel plate is used to attach to a female clip on the boot heel. A leather strap secures the toe area to the boot. Ninety percent of the original black paint still remains. Length 12", Width 2 3/4", Platform height 1 7/8", c. 1870. Class C: $100-200.

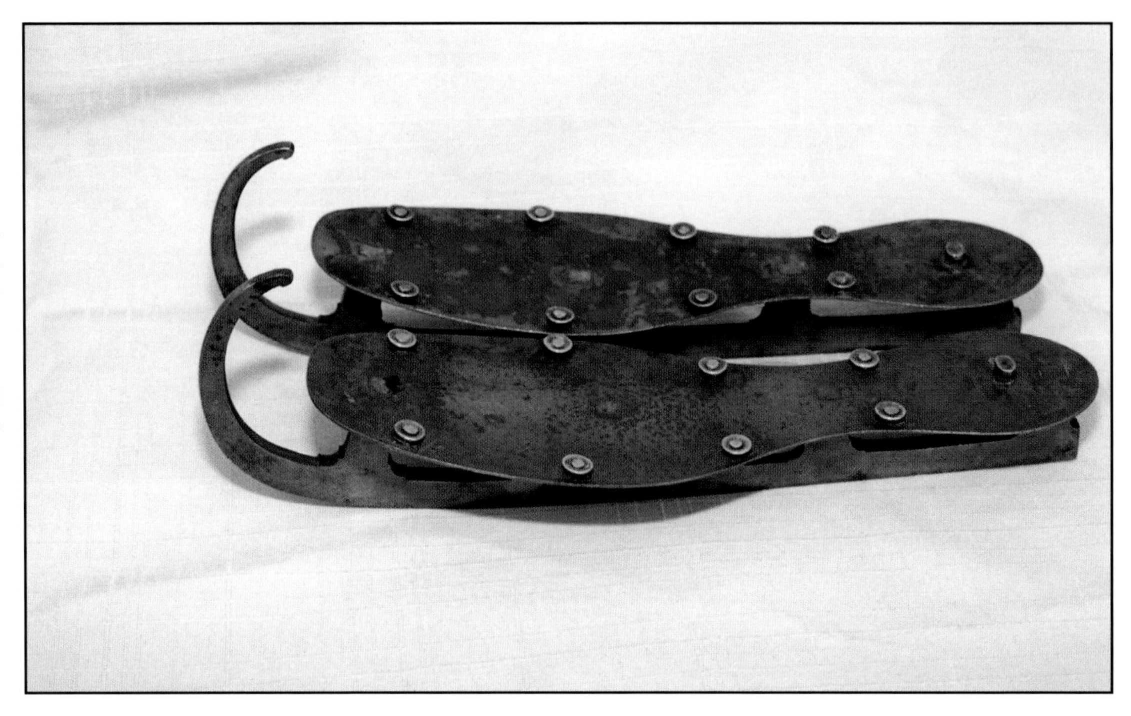

Figure 161. Factory made metal skates. The runners are forged and attached to the metal foot plate by four stanchions. The runner has a nice curl with a small finial at the tip. The four leather straps are riveted to the platform with copper rivets. The heel plate has a male knob that fits into a socket in the boot. Length 12 1/8", Width 3 1/8", Platform height 1 1/2", Curl diameter 3 7/8", c. 1860. Class D: $200-350.

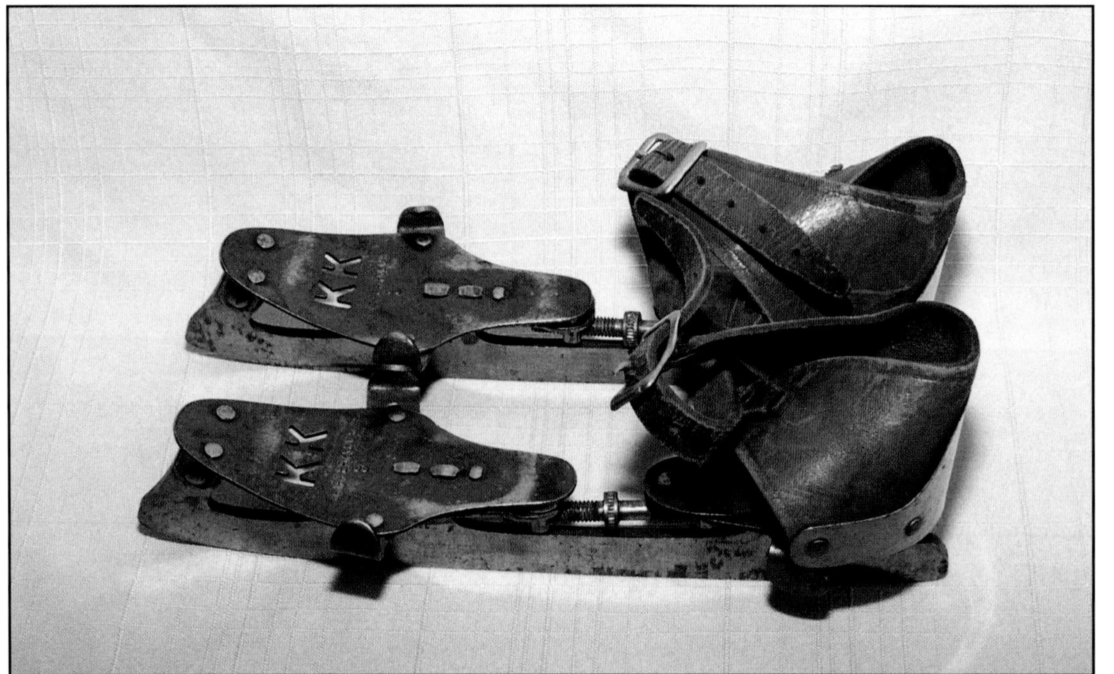

Figure 162. Women's metal club skates made by the Klipper Klub Manufacturing Co. The foot plate length and width is adjustable for different boot sizes. The heel cup leathers are attached by a riveted metal skirt. The skates are nickel plated and size 9. Length 10 1/4", Width 2 1/2", Platform height 1 1/2", c. 1880. Class B: $25-100.

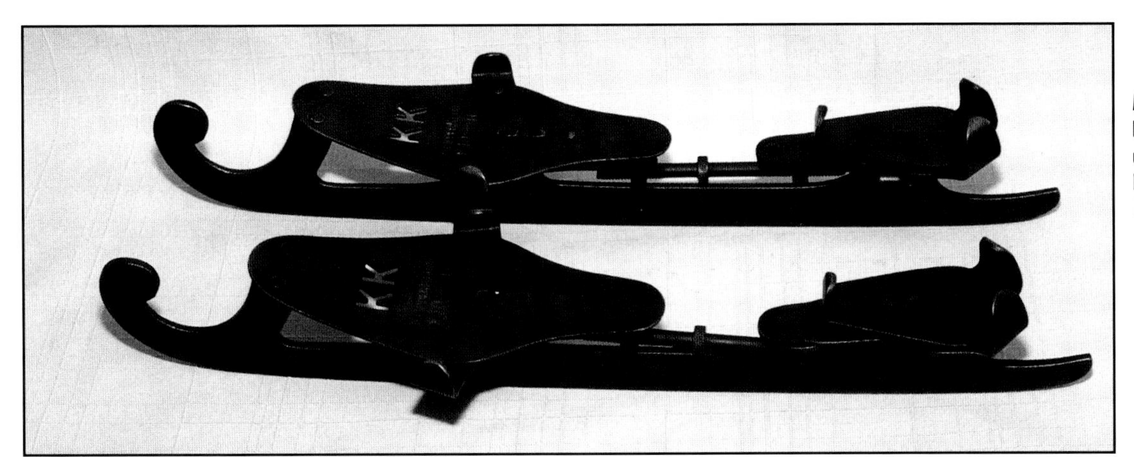

Figure 163. Long factory made club skates by Klipper Klub. The foot and heel plates are adjustable in length and width. The skate runner has an unusual curl and knob finial. They are stamped "F10-11 1/2, cast steel." Length 15", Width 2 3/4", Platform height 2", Curl diameter 2 3/8", c. 1870. Class B: $25-100.

Figure 164. Handmade skates with forged curls. The runners have two rectangular stanchion supports and nail spikes that extend through the heel. The wooden platform extending to the curl was painted red, much of which still remains. The platforms have two leather straps for fastenings. Length 10 1/4", Width 2 1/8", Platform height 1 1/4", Curl diameter 4 5/8", c. 1840. Class E: $350-500.

Figure 165. Unusual speed skates that are long and narrow. They are called the Torpedo model, manufactured by the Raymond Skate Co., Boston, Massachusetts. The platform is a round wooden dowel with a metal runner morticed into the bottom of it and secured with bolts. The front foot pad is oval and the rear is round, both made of metal screwed to the wooden platform. The long runner is stamped "warranted tool steel." Straps on the front hold the foot secure and screws on the heel hold the boot secure. Most of the old red paint still remains. Length 18", Width 2 3/8", Platform height 2", c. 1860. Class D: $200-350.

Figure 166. Handmade skates with beautifully forged and scrolled curls. The runner has two stanchions attached to the wooden platform with brass plates. The platform has sharp barbs on the foot area and spikes in the heel to support the boot. Two straps were used for fastening. The photograph was taken on a cold winter day south of Charm, Ohio, an Amish farming community. Note the old windmill on the hill. Length 11 3/4", Width 2 1/4", Platform height 1 5/8", Curl diameter 4 1/8", c. 1850. Class E: $350-500.

Figure 167. Handmade figure skates. They have large curls forged with brass finials on the tips. The forged runners have a curve on the bottom for figure skating. There are three leather straps for fastening them to the boots. The platform has an adjustable screw in the heel to secure the boot. The photograph was taken on a warm sunny day at the same place and on the same fence post as figure 166. Note the windmill wheel has turned 90 degrees since the winter photo. Length 12 1/2", Width 2 1/2", Platform height 2", Curl diameter 5 5/8", c. 1850. Class F: $500-750.

Figure 168. A graceful pair of handmade skates with large curls and brass acorn finials at the tips. The platforms were carved in an hourglass shape and fastened to the runner with a brass pin and square washer. Spikes on the heel supported the boot from sliding off the platform and two leather straps were used for fastening. Length 12", Width 2", Platform height 1 5/8", Curl diameter 5 5/8", c. 1850. Class F: $500-750.

Figure 169. This pair of skates has hand wrought grooved runners. The runner has a large curl on the front and forged finial at the tip. The walnut wooden platform is tapered down in the front to the runner blade. It has three leather straps for fastening and a large screw in the heel for supporting the boot. Length 13", Width 2 1/4", Platform height 1 1/2", Curl diameter 4 1/2", c. 1840. Class E: $350-500.

Figure 170. Metal skates in silhouette hanging from a flowering tree. The iron platforms are shaped like an hourglass and attached to the runner by three stanchions. The runners are forged into a nice curl. Leather straps are riveted onto the underside of the iron platforms. Length 14 7/8", Width 3 1/4", Platform height 1 1/4", Curl diameter 4 5/8", c. 1830. Class F: $500-750.

Figure 171. Extraordinary long racing skates. They are twenty-two inches long, the longest skates I have ever seen. This is a beautiful pair of factory made skates with a decorative brass plate on the toe in the shape of a cloverleaf. The cherry platforms are narrow at the toe and taper back to the wider foot plate. They have long screws in the heel for supporting the boot. Three leather straps were used for fastening. The wooden platforms have nice patina and color. Length 22", Width 2 1/4", Platform height 1 3/4", c. 1860. Class E: $350-500.

Figure 172. A spectacular pair of handmade Swan skates. They are exceptionally large hand forged swan's heads. The thick platforms have three leather slots for fastening and a hand forged spike on the heel for supporting the boot. There are two clips forge welded to the top of the runner that attaches to the platform. This is one of the finest pair of skates anyone would want to own. Length 11 3/4", Width 2 1/4", Platform height 2 1/8", Curl height 6", c. 1830. Class I $1500-2000.

Figure 173. Fine old handmade skates with a graceful curl. The finial was forged into an attractive diamond-shaped tip. The platform is supported by two stanchions and the runner at the front. The platform has sharp barbs on the foot area and a spike in the heel to support the boot. A nice harness cup arrangement was used at the rear for fastening the boot. Length 11 3/8", Width 2 5/8", Platform height 1 3/4", Curl diameter 5 1/4", c. 1840. Class F: $500-750.

Figure 174. A rare pair of swan's head skates. The hand forged runners are decoratively stamped with feathers on the head, neck, and breast. The swan's eye is made by a small hole drilled through the blade. There are three hand stamped bunny rabbits on the center of the runner with the letters MMOZ. The platforms are replaced with rosewood and a heel screw is used to secure the boot heel. Two leather straps were used for fastening. Length 12 3/4", Width 2 1/4", Platform height 1 3/4", Curl diameter 2 7/8", c. 1840. Class G: $750-1000.

Figure 175. Early factory made skates with tiger stripe maple platforms. The platform has brass trimmings to hold the leather toes and heel cup on. The toe and heel cup leathers are enclosed. The runners have small curls with finials at the tip. The platform rests on two brass stanchions. Length 10 1/2", Width 2 1/2", Platform height 1 5/8", Curl diameter 2", c. 1860. Class D: $200-350.

Figure 176. Handmade racing skates with turned over curls on the fronts. The rear of the runner has a decorative touch to it. The platforms are contour carved to the foot arch and cut down for the heel. They are attached to the runner by metal pins and brass washers. There are three morticed slots for the leather fastenings of the boot to go through. Length 13 7/8", Width 2", Platform height 1 1/8", Curl diameter 2 1/4", c. 1840. Class D: $200-350.

Figure 177. Early factory made skates. The runner is hand forged with a round finial at the tip. The front of the wooden platform rises up at the toe area. The runner is stamped "Klein's Patent 1188856-1857, Newark, N.J." The fastenings consist of leather straps at the toe and a harness with rings at the heel. Length 11 3/4", Width 2 1/8", Platform height 1 1/2", Curl diameter 4 1/8", c. 1856. Class D: $200-350.

Figure 178. Factory made skates with curls on the runners. There is a round finial forged at the tip. The platform has pins sticking up at the heel to support the boot. There are three metal slots on the underside of the platform to receive the leather straps for fastening. The runner is stamped "A. Tillmes and Co. Philadelphia, Pa. No. 6." Length 12", Width 2 1/2", Platform height 1 7/8", Curl diameter 4 3/8", c. 1860. Class D: $200-350.

Figure 179. An attractive pair of factory made skates. The runner has a beautiful curl and forged turn over tip. The platform has three leather straps to fasten the skates to the boot. The short heel screw for the boot has a brass washer around it. Length 11 1/2", Width 2 3/8", Platform height 1 3/4", Curl diameter 4 5/8", c. 1860. Class D: $200-350.

Figure 180. Early handmade skates with unique curls on the runners. The platform was painted black originally and is worn in the proper places. The platform has leather loops on the front and straps on the back for fastenings. The runner has two stanchions fastening it to the platform. Length 11 3/8", Width 2 3/8", Platform height 1 5/8", Curl diameter 4", c. 1850. Class D: $200-350.

Figure 181. A wonderful pair of handmade skates. The large curl on the front is hand forged into a tightly scrolled finial. You can still see the forged hammer marks on the curls. The runner blades are quite high and then cut down at the beginning of the curl. The platform has sharp pins at the heel and two leather straps for fastenings. Length 12", Width 2 5/8", Platform height 1 5/8", Curl diameter 4 1/4", c. 1830. Class F: $500-750.

Figure 182. A pair of long factory made skates with walnut platforms. The platforms have two barbs on the foot plate area and a long screw at the heel to secure the boot. They also have three leather straps for fastenings. The blade is stamped "Union Hardware Co., 'The Donoohue Racing Skate', Torrington, Ct." Length 16", Width 2 1/2", Platform height 1 5/8", c. 1866. Class C: $100-200.

Figure 183. An attractive pair of handmade racing skates. The runners are beautifully forged into scrolled curls and are made from old files, as the teeth marks can still be seen on the blades. A hollow or gutter groove is also cut on the bottom of the runner. The platforms are quite thick, contour-carved to the arch shape, and cut down at the heel. There are three leather straps used for fastening to the boot. Most of the original black paint still remains on the platforms. Length 16 1/8", Width 2 1/8", Platform height 1 1/2", Curl diameter 3", c. 1840. Class F: $500-750.

Figure 184. Women's handmade skates, with runners that are forged into a small curls on the fronts. The platforms are made of black walnut and carved into the shape of the hourglass. The platform has decorative brass skirts holding the leather straps and heel cups on. The leather straps are hand sewn with linen thread, as seen on the heel leathers. Length 12", Width 2", Platform height 1 1/2", Curl diameter 1 5/8", c. 1850. Class D: $200-350.

Figure 185. Decorative cast iron skates made by Robert Gibson Co., Birmingham, Ct. They are the Hawkin's Patent, March 28, 1865, and Shirley's Patent, April 12, 1861. The cast iron clips on the front of the skates held the toe secure and a threaded or adjustable dog at the back tightened up the heel. The heel adjustment pushes against two sharp spikes that hold the heel secure. The principle seems good, but I don' t know if it ever worked very well. Length 11", Width 2 3/4", Platform height 2", c. 1860. Class C: $100-200.

Figure 186. Primitive handmade skates with runners that were forged from old files. The runner blade has a little turn up curl on the front and a ring forged on the rear to screw to the platform. The platform is unusually thick and has the original red paint on it. It has sharp spikes on the heel and toe area to hold the boot secure. The leather straps on the heel, which were hand sewn with linen thread, combined with straps on the toe for fastening. Length 14 1/2", Width 2 1/2", Platform height 2 1/4", Curl diameter 1 3/4", c. 1850. Class D: $200-350.

Figure 187. Early hand crafted skates. The runner has a very bold and dramatic hand forged curl on the front. The platforms are fastened to the runner blades, with a screw at the heel area for securing the boot. Length 10 3/4", Width 2 1/8", Platform height 1 3/8", Curl diameter 1 7/8", c. 1860. Class D: $200-350.

Figure 188. Unusual metal racing skates perched on top of a fence post in the rural Amish country of Ohio. They are factory made with "M. Schmid & Sons, No. 29" stamped on the blade. The runner blade is thick and long with a large bold curl on front. The foot platforms consist of two cast brass pads and stanchions combined. The platforms have slots for the leather fastenings. The front oval casting has sharp spikes for securing the toes and large screws on the heel to secure the boot. Note the shocked wheat and four horses pulling a binder in the background field. Length 12 1/4", Width 2 3/4", Platform height 2", Curl diameter 2 1/2", c. 1860. Class D: $200-350.

Figure 189. This handmade pair of skates is perched on top of a wheat shock in the rural Amish country of Ohio. The hand forged runners have a nice curl with turn over finial. The platforms have screws in the heels and three morticed slots for the leather fastenings. Length 10 3/8", Width 2 1/4", Platform height 1 3/4", Curl diameter 4 1/2", c. 1850. Class D: $200-350.

Figure 190. Decorative handmade women's skates. The runner has a nice forged curl on the front with the tip doubled over, forming the finial. The oak platform has nice grain, color, and patina. The leather fastenings were nailed to the platform alternately by brass and metal headed nails giving the skates a nice appearance. Length 11", Width 2 1/4", Platform height 1 1/2", Curl diameter 2 3/4", c. 1850. Class D: $200-350.

Figure 191. An attractive pair of handmade figure skates. The runner blades were forged into big, beautiful curls with brass acorn finials at the top. The platforms have long screws on the heel to secure the boot. They have three leather straps through the morticed slots to fasten the boot. Length 12 3/4", Width 2 5/8", Platform height 2", Curl diameter 5 3/8", c. 1840. Class E: $350-500.

Figure 192. Large handmade skates with grooves forged along the runner bottoms. The runner has a large curl with a forged finial at the tip. The platforms are carved and tapered down to the curl on the front. The platform has a screw on the heel to hold the boot. Three leather straps were used for the fastenings. Length 12 3/4", Width 2 1/8", Platform height 1 1/2", Curl diameter 5 1/4", c. 1850. Class E: $350-500.

Figure 194. Representing blacksmithing at its best, this hand forged pair of skates was carefully thought out and skillfully constructed. The runner has a beautiful curl at the front and was attached to the platform with arched stanchions, front and back. The platform was formed by a triangular frame made out of flat bar stock, joining a round plate at the heel. The round heel plate has a sharp pin to hold the boot secure. The front foot is neatly held onto the platform with a round arched loop. All of the bars were forge welded in place with care and skill. Length 12", Width 4 1/8", Platform height 2", Curl diameter 4", c. 1860. Class F: $500-750.

Figure 193. A good example of a very long racing skate. It is an early factory made skate with an unusual front curl tip that is morticed into the metal top support plate. The top plate was screwed to the end of the wooden platform and then cantilevered beyond to support the runner curl. The back end of the platform had a support plate and long screw for securing the heel. After spinning the heel screw up into the boot heel several times the top of the wooden platform around the screw would get gouged up, so the heel plate was added to prevent it. The plates are nickel plated. Length 18 3/4", Width 2 1/4", Platform height 1 5/8", c. 1870. Class D: $200-350.

Figure 195. Outstanding, one-of-a-kind, handmade skates. The most unusual part of this pair is the double-ended curled runner. The rear runner curl is tucked up under the platform and the front curl is bold with a knob finial. The platform's wooden grain and patina is great. This is an example of the craftsman very carefully picking his timber! The leather harness fastenings on the rear are held together with copper rivets. The craftsmanship permeates from this pair of ice skates. Length 13 3/8", Width 2 3/4", Platform height 2", Curl diameter 2 3/8", c. 1840. Class F: $500-750.

Figure 196. This handmade skate is one of a kind. The runner has a nice forged curl on the front. The class of this skate lies in the beauty of the brass inlaid emblems. There are three brass top plates along with a diamond, anchor, and star emblems all inlaid smooth into the cherry platform. There is also a brass skirt attached around the edge of the platform to embellish the skate further. Perhaps this skate was made by a skilled seaman on a long voyage out to sea with a lot of time on his hands. The only sad thing about this skate is that it lost its partner years ago. If someone out there has its mate, let me know. Length 12", Width 2 3/8", Platform height 1 1/2", Curl diameter 4", c. 1840. Class E: $350-500.

Figure 197. These early hand crafted racing skates have a lot of class. The hand forged runner is great with the graceful curl extending beyond the platform and the decorative touch on the rear of the runner. The platform is contour-carved for the arch and notched down for the heel. The wooden platform was originally painted with a dark blue paint, much of which still exists. There are three morticed slots for the leather fastenings to attach to the boot. The leather straps were sewn with linen thread. Length 16 3/4", Width 2", Platform height 1 1/4", Curl diameter 2 7/8", c. 1840. Class F: $500-750.

Figure 198. The hand wrought runners of these attractive handmade skates have nice big curls on the front with forged finial tips. The wooden platforms still have their old red paint on them. There are two leather straps for fastening the skates to the boot. Length 12", Width 2 1/4", Platform height 1 3/4", Curl diameter 4 3/4", c. 1840. Class E: $350-500.

Figure 199. Factory made metal skates with the foot and heel plates drilled to screw to the bottom of the boot. It has the saw tooth rake feature on the curl. This is the Jackson Haines style from Sweden. The foot plate is in the shape of a heart and nickel plated. The skate is stamped "Salchow, Alb. Stille, Stockholm, Importe de Suede, 10 3/4." Length 10 3/4", Width 2 5/8", Platform height 1 3/4", Curl diameter 2 3/8", c. 1880. Class C: $100-200.

Figure 200. Metal skates with adjustable foot and heel pads for tightening. The heel pads have sharp barbs on them to penetrate into the heel sides. The skates are stamped "Kondor-Veloz, Stahlbahn, Gehartet, 25-27." The runners are nickel plated and have the saw tooth rake on the prow. Length 11 7/8", Width 3", Platform height 1 5/8", c. 1900. Class C: $100-200.

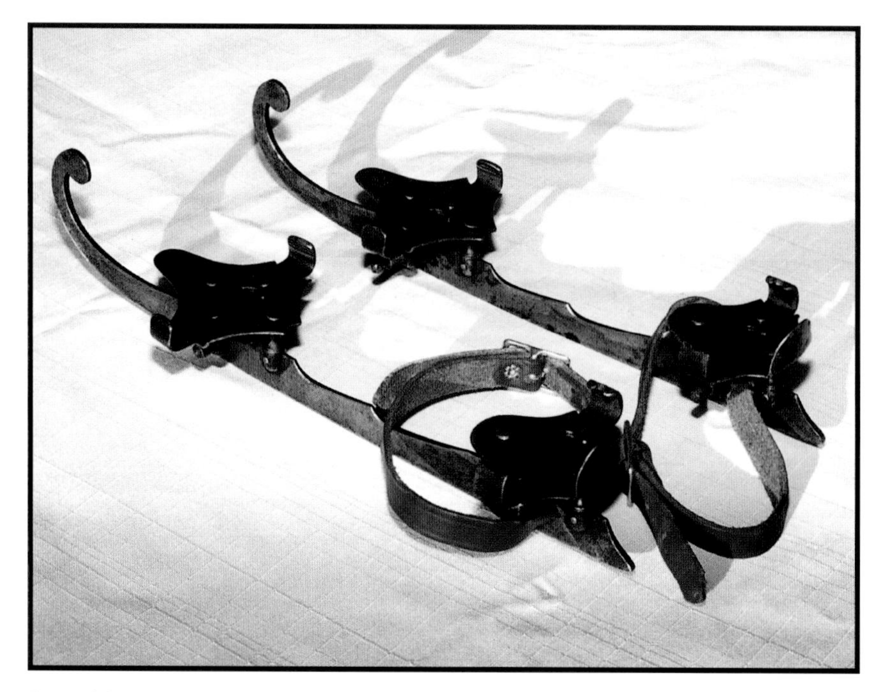

Figure 201. An unusual metal skate design, it has a dramatic curl on the runner and a nice decorative scallop on the top of the runner. There are two adjustable clips on the toe and heel for fastening. A leather strap is used around the ankle as well. It is stamped "Stahl" on the curl and "DRP 159375 Vorwarts 32" on the platform. Length 11 3/4", Width 2 1/2", Platform height 1 3/4", Curl diameter 2 3/4", c. 1870. Class C: $100-200.

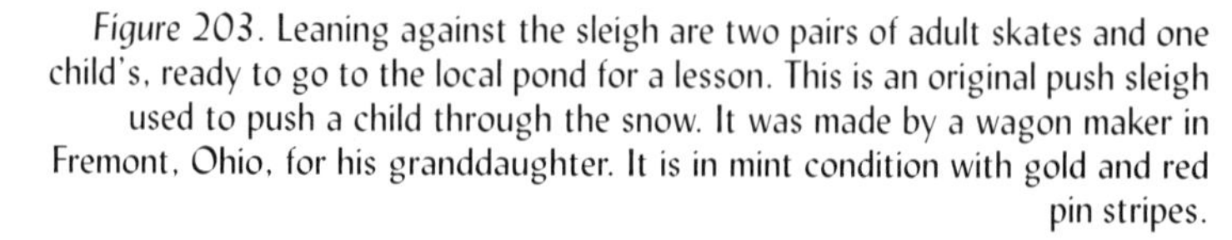

Figure 203. Leaning against the sleigh are two pairs of adult skates and one child's, ready to go to the local pond for a lesson. This is an original push sleigh used to push a child through the snow. It was made by a wagon maker in Fremont, Ohio, for his granddaughter. It is in mint condition with gold and red pin stripes.

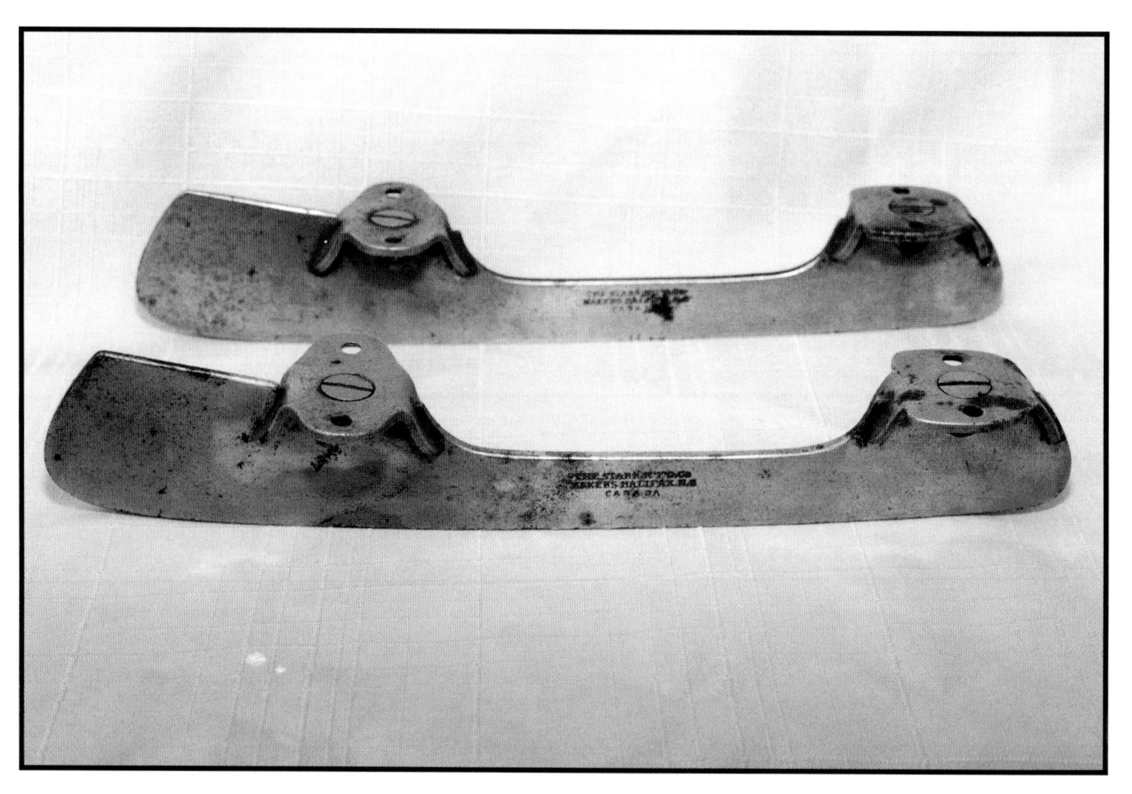

Figure 202. Metal skates that have two small platforms attached by screws. They were screwed to the boot for fastening. The skates were manufactured by the Starr Manufacturing Co., Halifax, Canada. They have the look of an old butcher blade. Length 10", Width 1 7/8", platform height 1 1/2", c. 1900. Class C: $100-200.

Chapter Six

American Patents On Ice Skates

During the 1800s many innovative individuals and skate manufactures made improvements to skate designs. They attempted to improve appearances and, more importantly, to ease the wearing and attachment to the foot. I can remember the old metal skates continually coming loose or falling off while skating as a youth. Many others can probably remember that same frustrating experience when growing up.

The new improvements in design were drawn up and submitted to the Patent Office in Washington, D.C. in a written document form called a "Letters Patent" which describes the new design in detail. The patent office reviewed the drawings and Letters Patent, and if found to be different enough from previous skate design patents, it was assigned a patent number under the inventor's name.

The Patent Office assigned consecutive patent numbers to all patent submissions on every subject. Patent No. 30195 was issued to J.F. Blondin, the inventor of an ankle support skate, on Oct. 2, 1860. See *Figures* 204-206 for a copy of the patent and example of the actual skates.

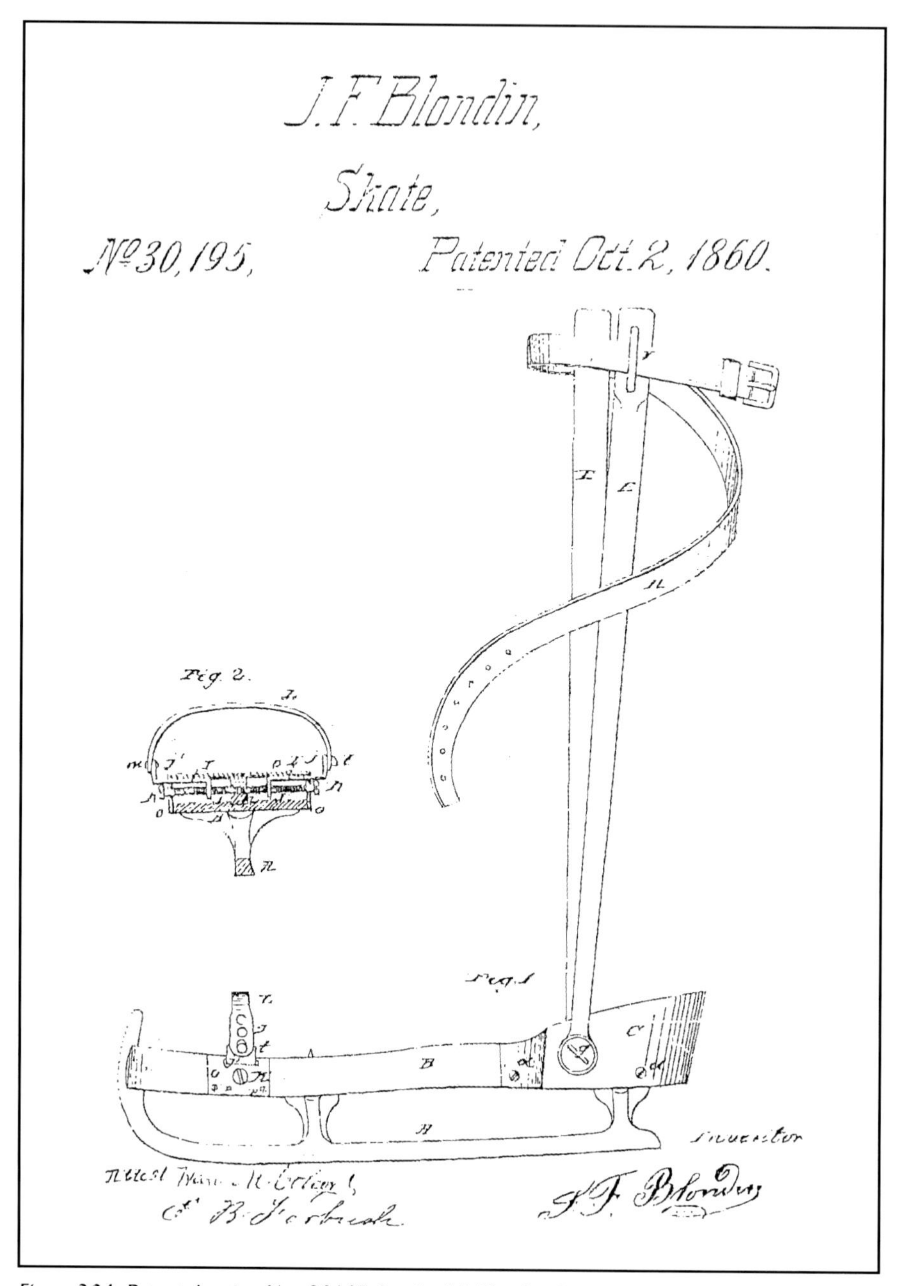

Figure 204. Patent drawing No. 30195 for the J.F. Blondin skate, patented Oct. 2, 1860. It illustrates the ankle support straps used on the skates.

Figure 205. A pair of J.F. Blondin skates in the flesh. They are a showy pair of skates with brass heel cups and support straps. All of the brass straps are decoratively engraved with a variety of designs and hand stamped with the maker's name, patent name, and date. The maker is the Douglas Rogers and Co., Norwich, Connecticut, who manufactured the skates under the Blondin Skate patent, Oct. 2, 1860. The front leathers were fastened to the platform skirt with brass straps. The platforms were made from maple with a heel spike for support. The leathers were originally a red color. The wealthy probably purchased these fancy skates. They are not too plentiful today and command good prices. Length 11 1/2", Width 2 7/8", Platform height 1 3/4", Curl diameter 2 1/4", c. 1860. Class F: $500-750.

Figure 206. The decorative engraving and patent date information stamped on the Blondin Patent skate.

If a skate manufacturer discovered a uniquely designed skate which might be manufactured and sold at a profit, the rights from the "patentee" would be purchased to manufacture it under the manufacturer's name. An example is the Douglas Rogers Co., who manufactured the Blondin skate under the Blondin patent described above.

Over time there were many new improvements on skate fastenings and attachments to the foot. Skates coming loose or off of the shoe were always a major problem, so designing the perfect solution to fastening the old skates properly to the shoes was always a goal. There are 52 different patents issued on skate fastenings alone from 1790-1873! *Figures* 207-214 are some of the unique and interesting patents on ice skates, illustrating several methods of fastening the skates to the boots, and a brief description of each.

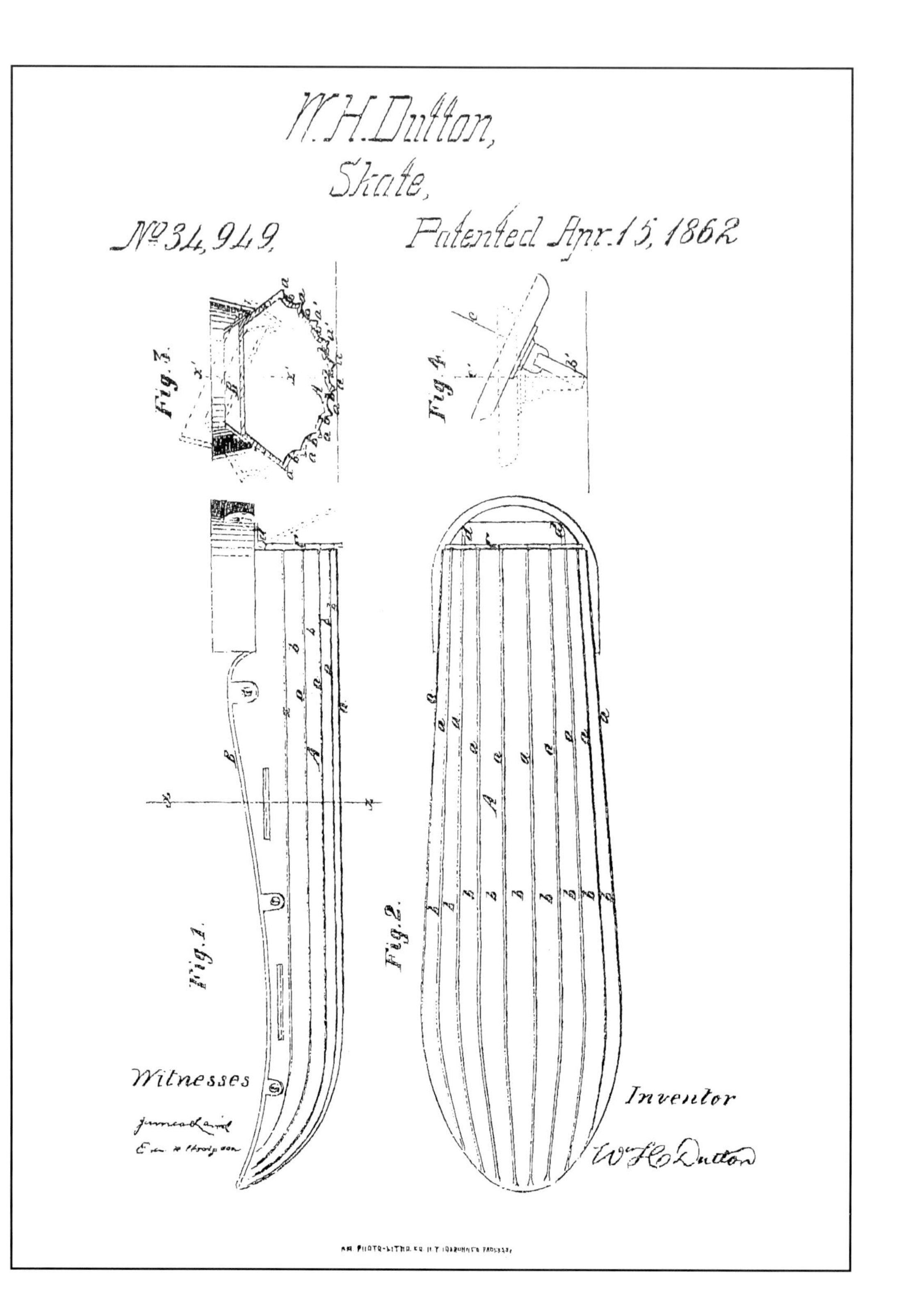

Figure 207. W.H. Dutton Patent No. 34949, Patented April 15, 1862. This skate patent has a set of skate runners all the way around the bottom of the skate to allow the skater to turn his ankle at any angle and still have a runner support on the ice.

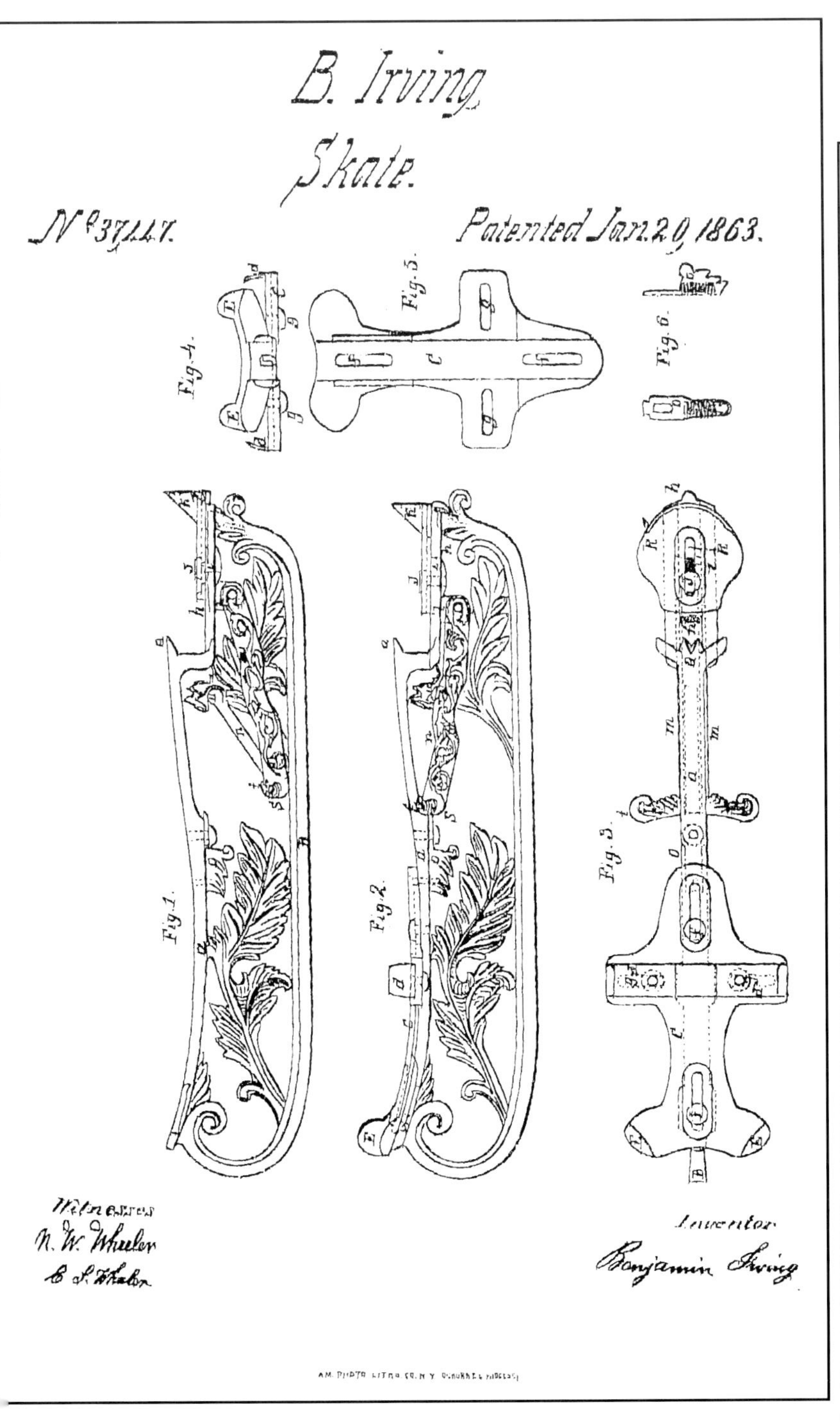

Figure 208. B. Irving Patent No. 37447, Patented Jan. 20, 1863. It is for an ornate, Victorian cast iron skate with a screw tightening device for fastening.

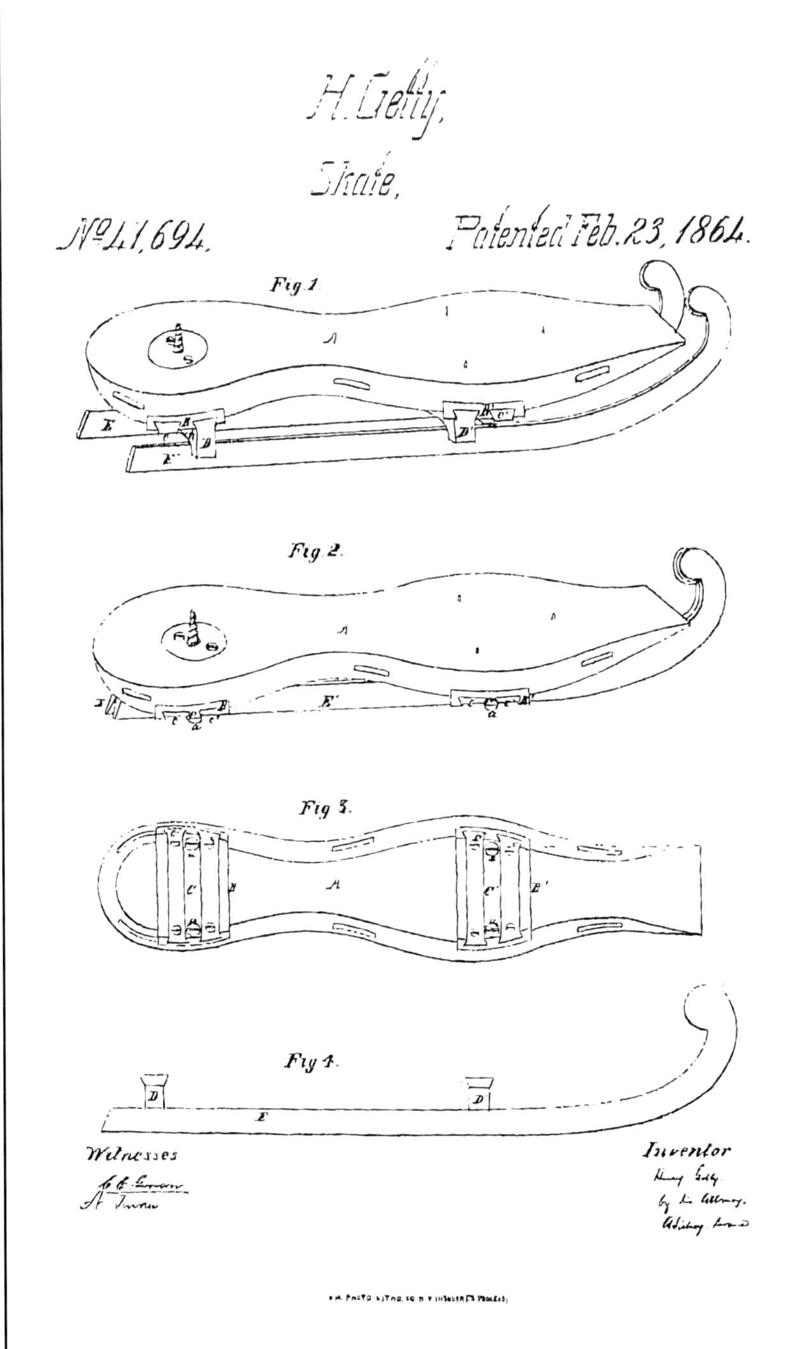

Figure 209. H. Getty Patent No. 41694, Patented Feb. 23, 1864. The Getty Patent had the option of a traditional single skate runner or a double runner attachment to the skate to accommodate the skater with more stability if needed. It could have a double runner for beginners and a single runner for the proficient skater.

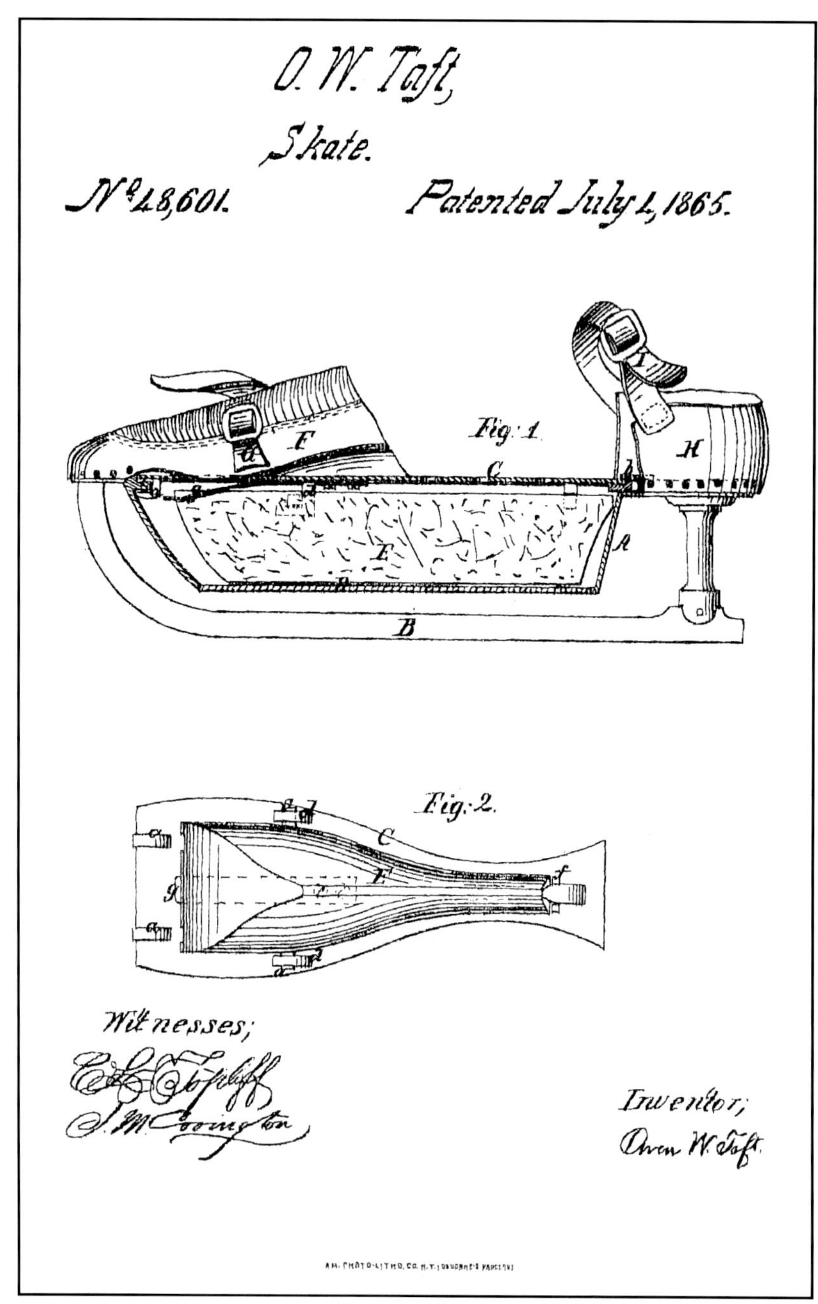

Figure 210. O.W. Taft Patent No. 48601, Patented July 1, 1865. One of the most unusual skate designs I have ever seen, it incorporates a hot iron slug as part of the skate platform to keep the skater's feet warm!

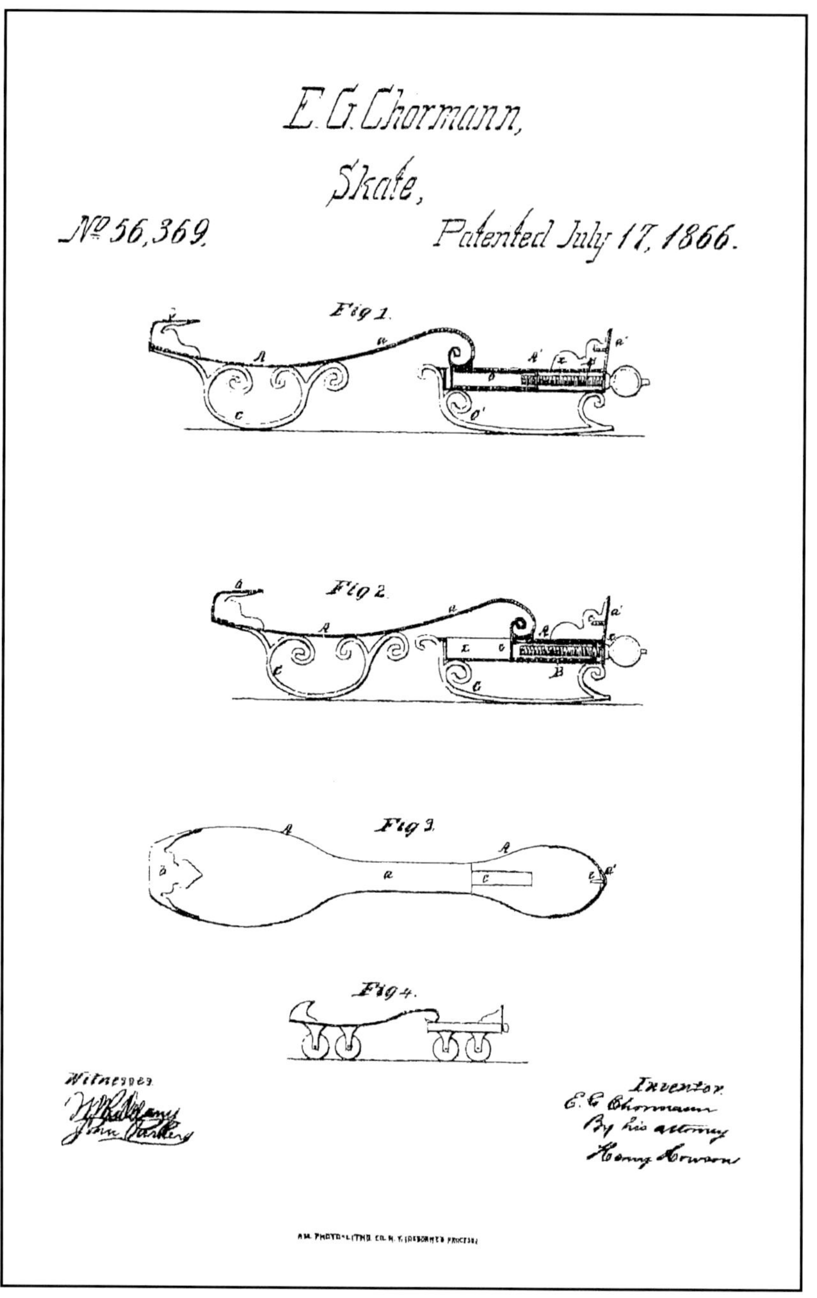

Figure 211. E.G. Chormann Patent No. 56369, Patented July 17, 1866. The Chormann skate patent was a very graceful looking ice runner design with an option to add rollers, making them roller skates! It looks a bit like a fancy bobsled.

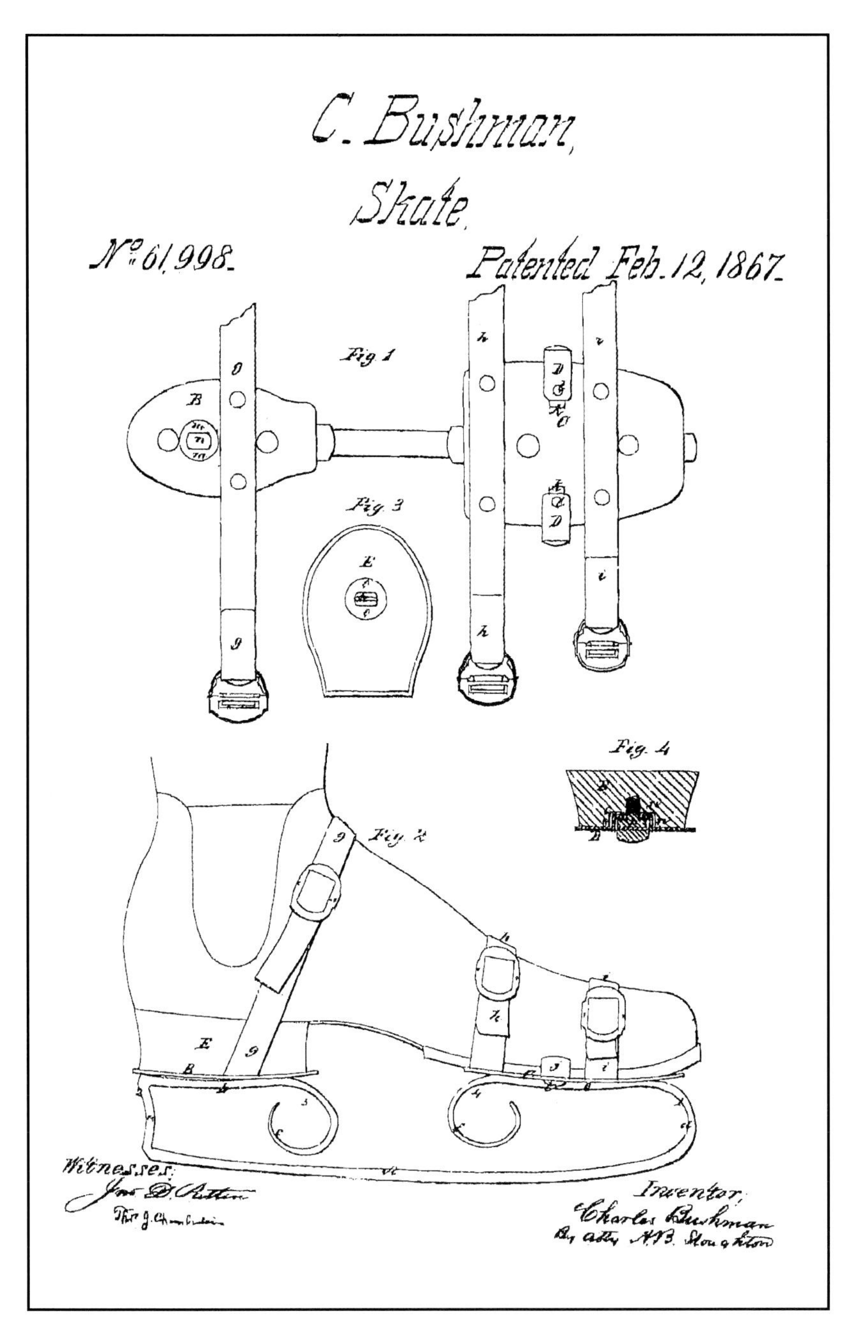

Figure 212. C. Bushman Patent No. 61998, Patented Feb. 12, 1867. The Bushmann design incorporated the runner and platform all in one piece, using the half coil runner to act as a soft spring action!

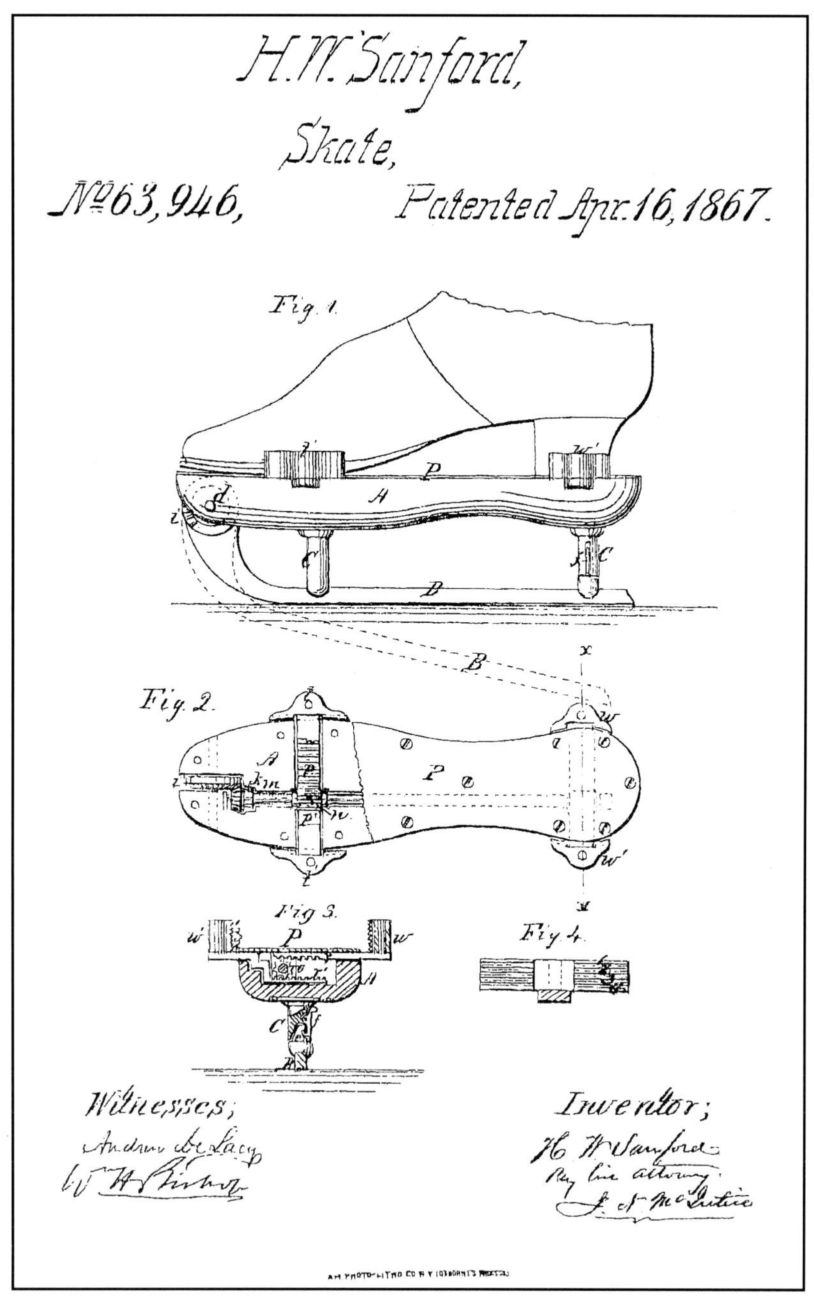

Figure 213. H.W. Sanford Patent No. 63946, Patented April 16, 1867. A unique rack and pinion gear mechanism clamped the skate to the boot when the top hinged platform came down onto the stanchion frame. It looks like the new clapp skate design!

Did any of these patent designs ever adequately address the problem? The dilemma was eventually solved when the upper shoe was permanently attached to the platforms and runners. An attempt was made in this direction by the Dutch skates shown with the uppers attached to the platform. It is difficult to date the transitional Dutch skate, but eventually all skates were made by riveting or screwing the runners permanently to the boot. This solved the centuries old problem of loose skates.

Skate development occurred at different times and places in America as well as in different countries. But each new skate development did not occur at the same time around the world, nor did all the countries accept these new design changes. Customs and traditions, like many other things, prevented changes in skate designs. But over time numerous patents were developed, including the ones previously listed, which certainly improved skate performance. Eventually skates evolved into the modern designs we use today, thus solving the problem.

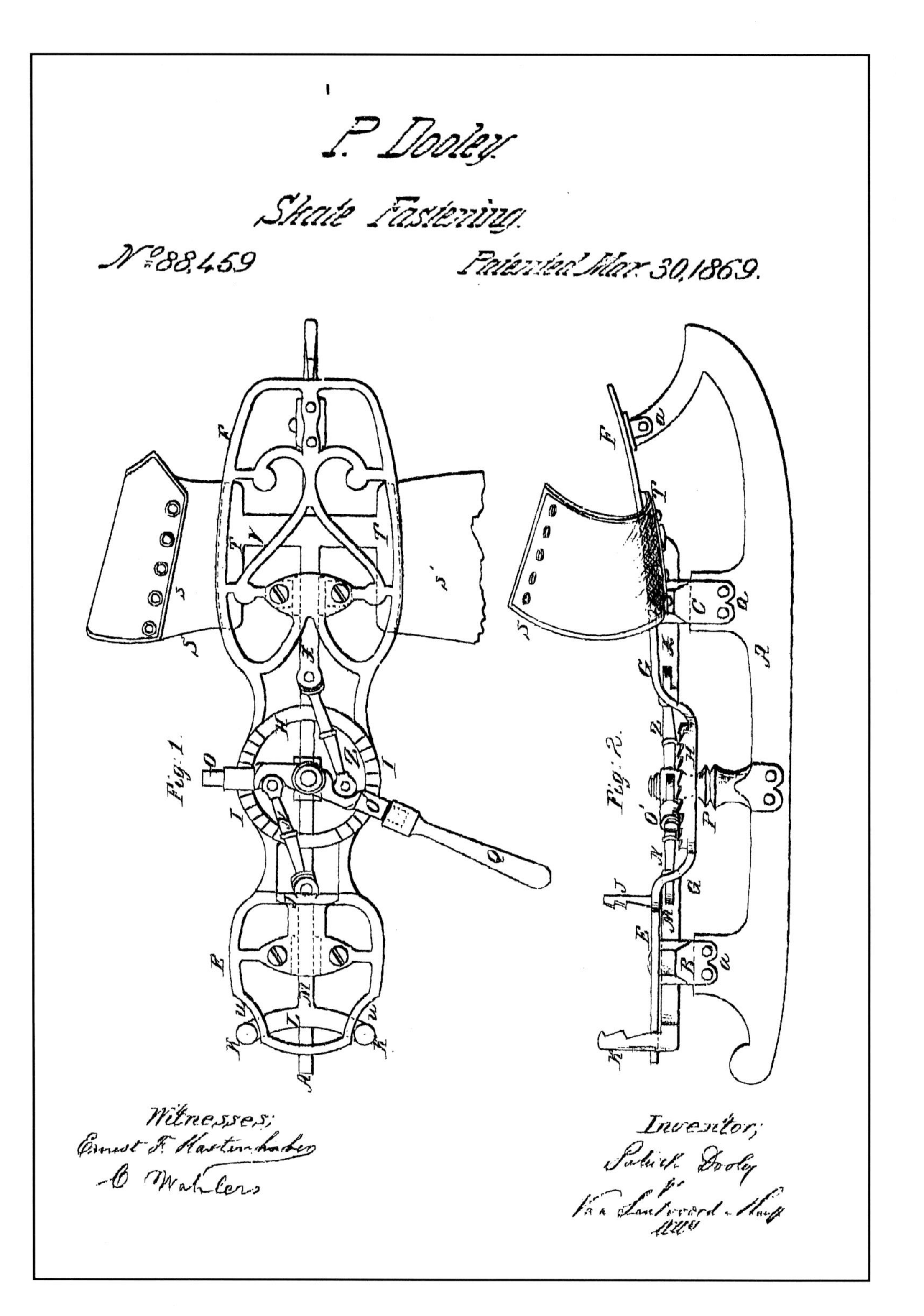

Figure 214. P. Dooley Patent N. 88459, Patented March 30, 1869. This design has a complicated ratchet and pawl tightening mechanism to clamp the skates tight to the boot.

Chapter Seven

Children's Ice Skates

Antique ice skates in general are quite collectable today, but the early children's skates are especially desirable. They are rare compared to the traditional adult skates, which makes them attractive to collectors, and people just like to collect "miniature" items in most other antiques. For these reasons the children's skates tend to command some high prices. A pair of child's handmade ice skates with a nice curl on the front is very unusual. Everyone seems to describe them as being very "cute." There are several nice factory-made child's skates in existence as well. See the examples of the children's skates accompanied by descriptions and explanations in *Figures* 215-228.

Figure 215. A nice little pair of children's factory made skates. They have decorative inlaid brass plates on the toe and heel area. The platforms are made out of walnut with the toe area thicker and elevated. The heel nut supports a long screw to enter the boot heel for support. There were two short barbs on the foot pad for support as well. The solid runner has a nice curl on the toe. The leather fastenings had wide straps on the toe and regular straps on the heel. Length 7 7/8", Width 1 7/8", Platform height 1 1/2", c. 1860. Class E: $350-500.

Figure 216. Small children's skates with a leather harness in the rear and leather loops on the toe for fastenings. A brass lock nut holds the runner to the platform. They appear to be Dutch skates. Length 8", Width 1 3/4", Platform height 1 3/8", c. 1920. Class C: $100-200.

Figure 217. Handmade children's skates. The runners are solid with a radius cut on the front serving as the curl. The walnut platform has a sharp spike to secure the boot heel. There are two sharp spikes on the foot pad of the platform to prevent slipping. Length 9 3/4", Width 1 5/8", Platform height 1 5/8", c. 1860. Class E: $350-500.

Figure 218. These handmade children's skates are old and unique. The hand forged runners have turn-over curls with brass acorn finials. The runner is curved on the bottom for figure skating. It appears that the tightly curled ends are part of the original runner design, rather than being bent over later. Length 9 1/2", Width 1 1/2", Platform height 1 3/8", Curl diameter 1 5/8", c. 1840. Class F: $500-750.

Figure 219. Children's metal skates made by the Barney and Berry Skate Co., of Springfield, Massachusetts. They are stamped "No.8." 19th century patent dates are Jan. 16, 66, April 8, 73, May 2, 76, Oct. 4, 81. The toe area adjusts tight against the boot by closing up the cam arm mechanism. The heel length is also adjustable with a lead screw key. Length 8 7/8", Width 2", Platform height 1 1/2", c. 1870. Class C: $100-200.

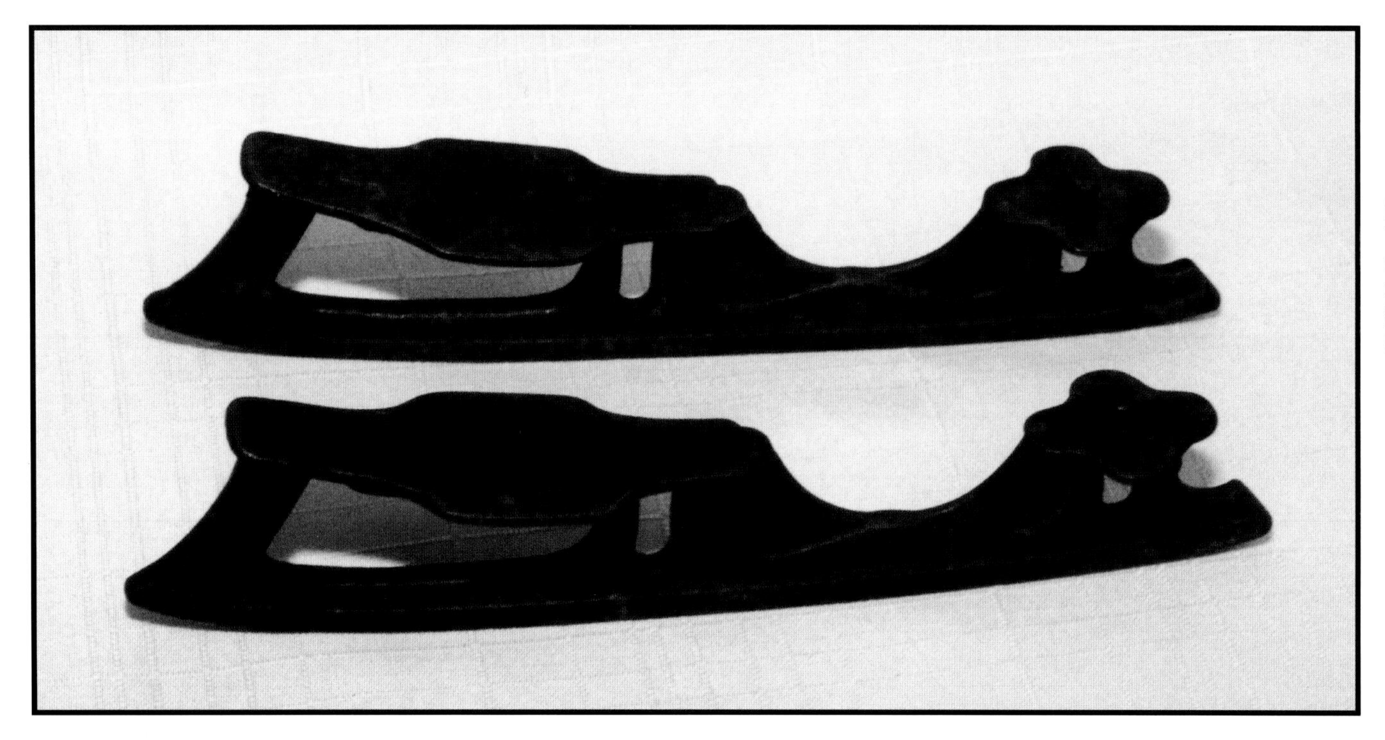

Figure 220. Cast iron children's skates. They are very small, narrow, and decoratively cast. A leather strap was used to fasten the toe area and a male pin that fit into the heel held the rear secure. Length 8 1/4", Width 1 1/4", Platform height 1 1/2", c. 1890. Class C: $100-200.

Figure 221. Outstanding children's factory made skates. The runners are nicely forged as part of the stanchions and then attached to the platforms by screws. The contour carved platforms were painted a dark red with ninety-five percent still remaining. The hand stitched leather harness fastenings at the heel and front are still in excellent shape. This pair of skates was probably owned by a wealthy family and well taken care of over the years. They are well made and rare. Length 8 1/8", Width 2", Platform height 1 5/8", c. 1860. Class F: $500-750.

Figure 222. A very small pair of handmade children's skates. The runners were morticed into the front of the wooden platforms. The platforms have exceptionally long screws at the heel for the boot. The foot pad has two nails sticking up to prevent slippage of the foot. The platforms have two straps for fastening. Length 7", Width 1 5/8", Platform height 1 1/4", c. 1860. Class E: $350-500.

Figure 223. Factory made metal children's skates. They are adjustable in length to accommodate a child's foot length. They were made by and stamped "The Samuel Winslow Skate Manufacturing Co. Makers, Worcester, Mass. USA Pat'd Jan. 15, April 29, 1900. Made in USA." The leather straps hold them to the boots. They look like miniature bob sleds. Length 8 1/4", Width 1 3/4", Platform height 1", c. 1900. Class C: $100-200.

Figure 224. A pair of metal, factory made children's skates. They were made by the Union Hardware Co., Torrington, Connecticut. The platforms are very high and have leather straps to hold them to the boots. Originally they were probably intended to be screwed to the bottom of the boot. They are double runner skates for the beginning skater. By looking at the rivets used, I believe someone fashioned these skates from two other metal skates. Length 8 7/8", Width 2 1/2", Platform height 2 1/4", c. 1910. Class B: $25-100.

Figure 226. This nice pair of children's skates may be factory made. The forged runners have nice curls on the fronts with brass finials. The acorn finials are rare on children's skates. The platform has an adjustable metal spike in the heel for the boot. Two leather straps were used for fastening them to the boots. There are traces of red-orange paint on them still. Length 9 7/8", Width 1 7/8", Platform height 1 1/4", Curl diameter 3", c. 1860. Class F: $500-750.

Figure 225. This is a first class pair of factory made children's skates. The hand forged runners have a nice curl on the front attaching to the platform. There are two brass bell-shaped stanchions that attach the runners to the platform. The platform skirts have a brass plate nailed to them, which sets them apart. The platform has two brass nuts holding the platform to the runner. Two straps are used to hold the skates to the boot. Length 8 3/4", Width 2 1/8", Platform height 1 5/8", Curl diameter 2", c. 1860. Class E: $350- 500.

Figure 228. These metal skates, factory made by the Union Hardware Co., Torrington, Connecticut, have the recognizable cloverleaf stamped emblem on the toe. They are stamped "No. 1624-8 Fine Quality Steel." The front toe and rear areas have adjustable metal clips to tighten them to the boot. Thousands of regular size skates were made but not many of these children's skates surface in the marketplace. Length 8 7/8", Width 2 3/8", Platform height 1 1/2", c. 1877. Class C: $100-200.

Figure 227. Small, factory made children's skates with curls on the fronts and knob finials at the tips. The runners attach to the platform with two metal stanchions and pins. They have a spike on the heel to secure the boot. They have two leather straps threaded through delicate cast loops to attach to the boot. They are stamped "Klein's Patent Newark, New Jersey 1856 & 57." I feel these children's skates are a bit rare. Length 8 1/2", Width 1 5/8", Platform height 1 3/8", Curl diameter 2", c. 1856. Class E: $350-500.

Chapter Eight

Ice Skates Accessories and Memorabila

There are endless ice skate accessories and types of ice skate memorabilia available. Only a few will be addressed to give the reader a flavor of them. Many of them have become collector's items themselves.

- Skater's costumes – men's, women's, Victorian.
- Hand and arm muff warmers.
- Tummy warmers, Dutch design copper belly tankard.
- Magazines illustrating early skating scenes of skaters in action such as the *Harper's Weekly*.
- Memorabilia and records of the skating winners in local, state, national, international, and Olympic skating competitions.
- Ice skating medals, medallions, and trophies.
- Newspaper clippings of historic ice skating events.
- Paintings, engravings, skate advertising posters, post cards, and photographs of ice skating scenes.
- Ice skating printed programs from national and international contests, shows, and exhibitions.
- Olympic pins.
- Printed cardboard posters of ice skating events.
- Sheet music for skating in the ice rinks.
- Ice skating related jewelry for collectors.
- Autograph collections of famous ice skaters, such as Olympic stars.
- Films and tapes of skating exhibitions, shows, and contests.
- Dolls attired in ice skates and ice skating costumes.
- Antique skater's lanterns.

Figure 229. This is an ice skater's accessory, a "tummy warmer." It is a solid copper canteen that is concave or circular to fit the belly. Note the brass cap at the top for filling it. There are two metal loops at the top for leather straps to hang around one's neck. It was filled with warm oil or sand, which would stay warm for awhile, and could be heated up again near a fire. Water was not used as it would freeze and destroy the warmer. It was purchased in Holland and used there. Length 10 3/4", Height 6 7/8", Thickness 9 /16", c. 1880. Class C: $100-200.

Figure 231. A color postcard with birthday greetings sent by a mother. It illustrates a young girl slipping down with a box full of good luck charms falling her way. Note the curled skates she is wearing.

Figure 230. A Barney and Berry advertising cardboard dated 1917. It illustrates the variety of metal skates they manufactured during the early 1900s. Note the prices of the various skates.

Figure 232. A colorful postcard illustrated with a nice young lady with a fancy hat ready to go skating. A pair of figure skates is hanging around her neck.

Figure 233. This winter scene 1909 postcard captures six skaters getting ready to start a race by the firing of the gun.

Figure 234. This is a very colorful scene illustrating two young girls ice skating and a boy on his way down. There is a lot of action in the painting, with the hat flying as he hits the ice. Wm. A. Bald & Co., Schaefferstown, PA.

Figure 235. An advertising card for Edwin C. Burt Fine Shoes being sold in Providence, R.I. Copyrighted by A. B. Sealy 1881.

Figure 236. A beautiful advertising card featuring two skating young women. The card is promoting Dean Bros. Boots, Shoes, and Rubbers, Taunton.

Figure 237. A colorful postcard with three young boys skating on a pond. It is a Christmas card embellished with holly and decorative designs.

Figure 239. An old advertising card with a pair of colorfully dressed-up young skaters. The card is advertising the Ivers and Pond Piano Co., Boston, Mass.

Figure 238. A neat old Christmas card illustrating a pair of skaters on a pond. Note the Christmas greenery and bird encircling the skater's frame.

Figure 240. Skating in Central Park, New York, is illustrated in this old postcard. Note the fancy dresses worn by the ladies on the right.

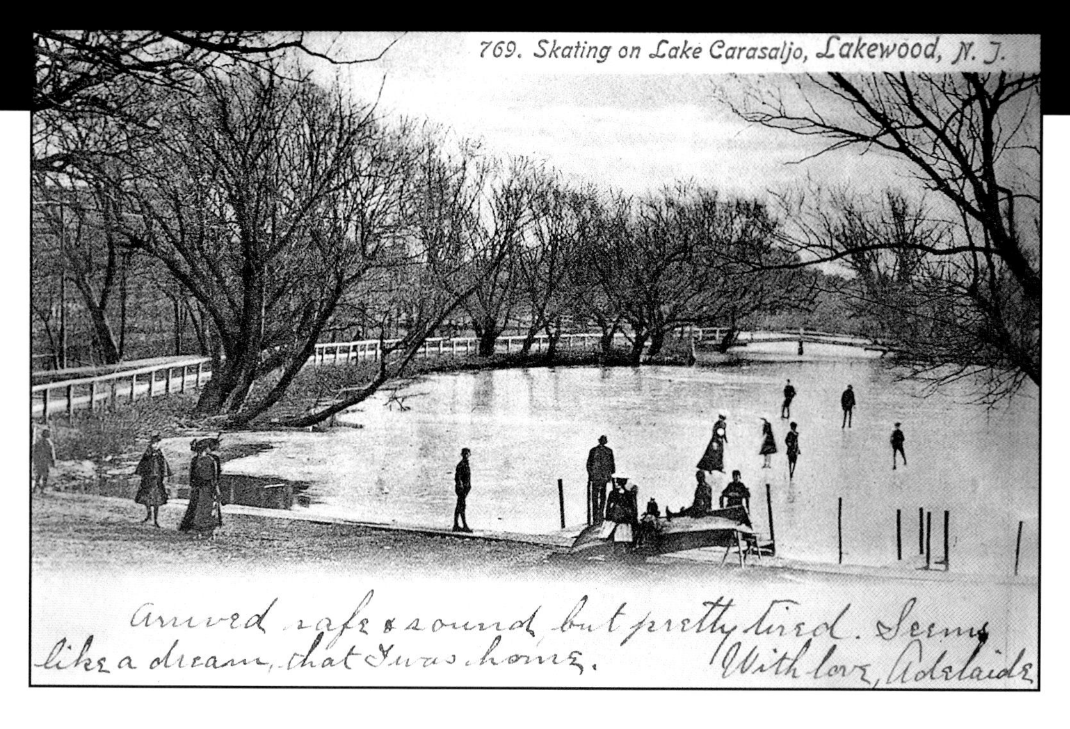

Figure 241. A few skaters enjoying themselves on Lake Carasaljo, Lakewood, New Jersey, and shown on this old postcard.

Figure 242. This old black and white postcard shows the action of a speed skater leaping over six wooden barrels. It seems to be a risky and dangerous sport to me! Note the men in the background with the brooms to clean the snow off the ice. G. T. Rabineau, Lake Placid, New York.

Figure 243. The skating rink at Shattuck Inn, Jaffrey, New Hampshire is captured in this old photograph. The skaters are more interested in getting their picture taken than wanting to skate. Note the big Shattuck Inn located in the background.

Figure 244. Glory of a Winter Day, by A.B. Frost, c. 1900.

Chapter Nine

Skater's Lanterns

In the past when the ice skaters went out at night, they carried small skater's lanterns over their wrists to help them see. These kerosene lanterns were attached with a small wire bailed handle. They could then see a little better at night while skating on the lakes, ponds, and rivers. Other skater lantern designs had a short chain attached to the top of the lantern with a small finger loop to hold it.

Though skaters would often have fires started on the banks of a river or pond to see by, warm up with, and congregate around, the skater's lantern provided light when the fires were far away, not available, or extinguished. It would be a big problem skating down a river on a partially moonlit night and then accidentally run into a log sticking up through the ice or into another skater!

The skater's lanterns were made in different sizes and designs but the most common one is shown in *Figure* 245. Several companies manufactured this typical lantern design in a variety of materials and finishes. The lanterns were made of tin, tin nickel plate, tin painted, solid brass, and tin and brass combination.

Figure 245. A traditional brass kerosene skater's lantern. Note the brass bail handle for carrying it and the stem sticking out for adjusting the wick height. Lantern height without bail handle 6 3/4", Tank base diameter 3 1/8", c. 1860. $100-250.

Most of the globes of the standard design had clear glass and most were approximately the same size and shape, but some designs varied slightly in diameter, shape, and size by the manufacturer. The much rarer globes were made in a variety of colored glass, including red, blue, green, cobalt blue, amber, and purple. The colored globes were manufactured in various combinations with tin, brass, and copper bodied lanterns. The colored glass globes with the brass body lanterns are quite rare and in demand. They command high prices and seem to be going up each year.

There are many other unique designs, styles, and shapes in skater's lanterns available as well. Many have uniquely shaped globes and lantern bodies, which are much rarer than the traditional style. Most of the lanterns have a small kerosene container on the bottom with an adjustable wick that delivers the fuel up to the top to be burnt. The wick height or amount of flame can be adjusted by turning a small stem sticking out at the base of the globe. There are also a few skater's lanterns around that have candles mounted in them. A variety of skater's lanterns with different shapes and features always complement one's ice skate collection.

To my knowledge the skater's lantern is an American invention. I showed photos of skater's lanterns to several people in museums in Holland and they all said they had never heard of or seen them. They thought the skater's lanterns for night skating was a good idea. I wonder what the European skaters used at night to see while skating. Maybe they never skated at night.

A collection of diverse skater's lanterns is illustrated in *Figures* 246-252.

Figure 246. An unusual skater's lantern with a brass cone top and copper tank base. This quality lantern is carried by a finger loop on a brass chain. A brass tag on the lower copper tank is embossed: "Directions, use only kerosene oil with the No 1 cyclone & convex burners." The stem reads "Manhattan Brass Co., N.Y." The brass cone is stamped "Patented Dec. 24, 1867." The molded glass globe reads "Pat. 5 April 1864." Lantern height 9 3/8", Tank diameter 3 1/2", c. 1860. $400-600.

Figure 247. A triangular frame and flat glass panels make this an unusual skater's lantern. The manufacturer is Holmes Booth and "Haydens Waterbury, Connecticut," is embossed on the stem wheel. The tank base is made out of brass. The triangular frame slides up the two bail wires allowing one to light the wick. Lantern height without bail 9 3/8", Tank base diameter 3 5/8", c. 1860. $400-600.

Figure 248. A brass skater's lantern with a dark green globe. The dark green is not too common on the market. Lantern height without bail 6 5/8", Tank base diameter 3 1/8", c. 1860. $500-700.

Figure 249. A beautiful brass skater's lantern with a cobalt blue globe. Several shades of blue were manufactured, but the cobalt blue is the most striking and appealing to the collectors. The cobalt blue is not found too often. Lantern height without bail 6 5/8", Tank diameter 3 1/8", c. 1860. $600-800.

Figure 250. An attractive brass lantern with a rare purple globe. These lanterns are not too plentiful or easily found. If your girlfriend was skating with a rare purple lantern at night you could find her easily. Lantern height without bail 6 7/8", Tank base diameter 3 1/8", c. 1860. $800-1000.

Figure 251. This group of rare skater's lanterns illustrates a variety of colored globes. This group includes cobalt blue, three other shades of blue, red, two greens, purple, and a clear globe for comparison.

Figure 252. These skater's lanterns illustrate the variety of designs, shapes, sizes, and different materials they were made out of. This group of lanterns is by no means all inclusive of all the designs made, but it will at least whet your appetite on the subject. These miniature kerosene lanterns are cute and collectible!

Opposite page

Figure 253. These are two pairs of hand carved ice skates made by the author. The platforms of the larger pair are made of burl walnut and the curled runners are carved out of one piece of tiger stripe maple. The platforms of the miniature pair are made of tiger stripe maple and the curled runners are carved out of wild cherry. The miniature skates are only 6 3/4 " long. This was a very challenging project!

Chapter Ten

Collector's and their Collections

Reasons for Collecting Ice Skates

There are as many motivations for collecting as there are people who collect. I have ten that I think are most common, but one may collect them for a combination of these reasons or reasons all their own.

1. Skates may bring back pleasant memories of skating as a youth.
2. One may appreciate the craftsmanship required in making an early pair of skates.
3. One may enjoy the graceful designs of the beautiful curled skates as a work of art.
4. One may enjoy skates merely as an old antique.
5. Some may appreciate them as a part of our early American heritage and family sport.
6. Others may enjoy looking at them hanging over their fireplace mantle on a cold winter evening or during the holiday season.
7. Others may enjoy them because they are accomplished ice skaters themselves in figure skating, speed skating, or hockey. They could be competing in state, national, or international competitions and just want a few old pairs of skates to remind them of their predecessors.
8. One may collect antique ice skates because they are just a unique collectable and people often admire them.
9. One may collect them purely as a good investment, sometimes better than the stock market.
10. One may be an avid fan of the ice skate competitions held internationally and may have always wanted an old pair of skates worn by the earlier competitors.

Displaying Ice Skate Collections

The following are some suggestions for displaying skates, particularly during the winter months and the holiday season.

- Businesses can display them on the walls of reception or waiting rooms for the pleasure of the customers or clients.
- Restaurant owners can hang them from the walls, above the bar, or in the entrance foyer.
- Historical organizations can display them at any of the historical buildings or mansions in town on a temporary or permanent basis.
- Members of a civic or social club could arrange for a table display at a luncheon or dinner during the winter or holiday season. One might consider putting on a short program talking about the collection.
- An antique wooden sled could accentuate a couple of pairs of old skates. The skates could be placed on the sled's platform along with a skater's lantern or all could be hung up on a wall for a display.
- Several pairs could be hung from the fireplace mantle or set on top of the mantle, along with some pine greenery for the holidays.
- A pair could be hung from the stairway spindles going upstairs in a home.
- A couple of pairs could be placed alongside a skater's lantern on a deacon's bench in the foyer entrance, as though people just came back from a skating party on the local pond.
- Several pair could be hung on a hand hewn wooden beam along the wall and ceiling of the kitchen, as I have done.
- Several pair could be placed on a shelf with the possibility of back lighting to enhance them.

- A pair adorned with red ribbon and greenery could be attached to the inside door during the holidays for friends and family to see.

I hope these suggestions will offer a few ideas on how to display collections to friends, neighbors, and the public. Happy displaying!

Cleaning Methods and Preservation of Old Ice Skates

Carefully remove the old leather fastenings if they are not permanently attached. Make note of their orientation through the skates or attachments. If the leathers are attached permanently be very careful not to tear them while cleaning the other parts of the skates, as they are often very dry and brittle.

Examine the blades very carefully before any cleaning is done to determine if there are any stamped or engraved manufacturer's names, patent names, or dates visible. Many times with a surface coating of rust they are not very legible or in many cases they are lightly stamped and can be easily overlooked. Examine the wooden platforms as well, since they were stamped occasionally. The maker' s stamp can be destroyed unknowingly if care is not taken.

CAUTION! DO NOT put the iron skate blades on a hungry, powered wire wheel to clean off the rust patina down to the gray iron! The value and looks of a $200 pair of ice skates can be quickly destroyed in the matter of a few minutes. In fact I will not buy a pair of skates destroyed by overcleaning unless it is a rare pair.

Following are a few suggestions on cleaning and preserving old ice skates.

Blade runners and metal parts of the skates. For metal blades that have a thin coating of rust on them use a fine "000" steel wool to loosen the rust and then wipe with a soft cloth. When the surface rust has been removed leaving a nice patina, apply a preservative coating of some type.

If the rust is very heavy a "00" steel wool or a very fine 220 or 320 grit sandpaper must be used. Use care with the sandpaper, avoiding scratches in the blade, and then follow up again with the "000" fine steel wool. Afterwards a coating of thin oil should be applied to the metal runners of the skates and then wiped dry.

A coating of clear Antique Oil Finish (soft luster) made by Minwax may be applied on the metal runners. This leaves a finish that lasts and seems to prevent further rusting. But if this product is used, apply it with a rag and wipe it off immediately with a dry soft cloth and let dry. Do not let it set on the metal runners too long, as it will get tacky. If it does get tacky just apply a little more fresh oil and wipe it dry again. After 24 hours buff the metal runners with a soft cloth to a smooth soft luster. The same oil works super on the wooden platforms of the skates as well. It leaves a nice patina finish on both wood and metal.

Wooden parts of the skates. Many times the old wooden platforms will have the original paint on them or varying degrees of paint, such as traces of red, green, blue, etc. In all cases do not remove the remaining old paint; try to preserve it since it adds value to the skates.

Carefully use "000" fine steel wool to clean the wooden platform, both painted and unpainted surfaces. If the old paint has a nice patina and color do not touch it. If the platform has surface dirt on it, go over it lightly with fine steel wool to remove the surface dirt and bring the traces of old paint back to life. Then apply a coat of clear Minwax Antique Oil Finish and wipe dry. After 24 hours buff smooth with a soft cloth. It will bring the wood grain out and brighten the old paint colors. The skates will look terrific. A Danish oil finish or one of your own finishes may be used as well.

Do not touch a well cared for pair of skates such as Victorian skates with walnut or rosewood platforms with the original varnish finish on them. Only clean the surface dirt with a damp cloth followed by a dry one and possibly apply a coat of paste wax and buff. Use good judgment in cleaning skates, and when in doubt do not do anything to them. Leave them as they are.

Leather straps and fastenings. In most cases an old pair of ice skates will be missing the original leathers or they will be in bad shape. It is important to preserve the old leathers if they are reasonably complete. If a leather or saddle shop is nearby, the straps can be replaced.

If the skates have the original leathers on them, attempt to preserve them with care, as they will no doubt be very dry and brittle. Preserve them by applying leather conditioners and cleaners. The liquid types, some with spray attachments, seem to soak in and work better than the paste types. Paste types require rubbing and working the preservatives into the leather. The bending and manipulating of the leather straps may cause them to break or come apart. If the leathers seem to be very brittle and weak it is better not to do anything with them. They will probably fall apart with handling so leave them as is. Leather preservatives can be purchased from a leather or saddle harness shop that specializes in leather goods.

If the seller is uncomfortable about cleaning the skates properly, leave the cleaning and proper repairs to the collector. As in antique tool collecting, the collector usually wants to clean and preserve the tools himself. They are more valuable untouched to the collector than improperly cleaned by the seller.

Recently I observed a nice pair of ice skates in a museum that had been overly cleaned right down to the gray iron. Some type of acid method or naval gel was apparently used to clean the iron parts. In my opinion they looked just horrible and the cleaning destroyed the value of the skates. Even some of the museums over clean their artifacts.

Chapter Eleven

General Price Guide For Ice Skates and Skater 's Lanterns

Skates

This book is intended to provide a general guideline to the collector, dealer, and any other interested person in antique ice skates. Following is a categorization of the various skates into classes by design, style, rarity, age, and general price range. An index code can be compared to each photograph and description of each skate. The price index code is based upon a pair of skates that is 100 percent complete, including the original leathers, with nothing chipped or cracked, and having only normal signs of wear and tear.

As a collector I could never find any guidance on prices for antique ice skates or any estimated values in antique price guide books to help me in my purchasing of skates. The only price guide on buying ice skates is one's own good judgment, or the dealer's tagged price which varies considerably. This price guide below is ONE person's opinion. Each collector will have his/her own mental guide on price, rarity, and skate design preferences. The price of antique ice skates will vary from one part of the country to another as well. Like any other antique, price is established and depends on the quality, condition, and a negotiated price between a buyer and a seller. This price guide is only a starting point.

Questions that may be of assistance in determining the ice skates value include:

- Are there any missing pieces on the skates?
- Are there any chips or flaws in the wood?
- Are the blades broken or cracked anywhere?
- What kind of wood are the platforms made of?
- Are the original straps still attached?
- Are the straps broken, brittle, or in good shape or hand stitched?
- Are the heel screws missing, damaged, or identical?
- Did someone over clean the iron and wood or put some kind of sticky brown lacquer on them?
- Are there any decorative brass inlays on the platforms, any brass embellishments on the tips of the runners, any stampings or emblems, any manufacture's name visible, or any patent dates shown?
- Is the original paint still visible on the platforms and worn in the proper places, or are they newly painted because someone thought it would enhance their value?
- Are they deeply pitted with rust or do they have typical rust deposits considering their age?
- Are they nice and shiny?
- Are the platforms toggled to the runners with cadmium plated Philips head screws, or are they handmade screws fastened in the proper manner?
- Did some wood butcher attempt to improvise the original design of the platform and destroy its integrity?

Obviously, the rarity and market value of an old pair of skates depends on about 50 different things that MAY or MAY NOT be important to you as a collector or buyer. Again I want to emphasize that this guide is to be used only as a general base and the values are NOT carved in stone. VALUE is only in the eye of the beholder.

MARKET VALUE CODE

Class A Less than $25
Class B $25 to $100
Class C $100 to $200
Class D $200 to $350
Class E $350 to $500
Class F $500 to $750
Class G $750 to $1000
Class H $1000 to $1500
Class I $1500 to $2000

Examples for each dollar value class includes SKATE DESIGN, RARITY, AND A GENERAL DESCRIPTION.

CLASS A Less than $25.

1. Common (all metal) clamp on skates with rust on them.
2. Common Skates, either all metal or combination of metal and wood, not complete or with modified platforms or leathers.

These skates are common and found at flea markets and lower quality antique shows. Most of the old skates found fall into this class.

CLASS B $25 - $100

1. Unusual metal clamp on skates in excellent condition.
2. Combination metal and wood platform skates with straps and no curl on the runner, and no character.

CLASS C $100 - $200

1. Wood or metal skates with small curl and straps.
2. Early patented skates complete with some rust.
3. Handmade skates of lesser quality and no straps.
4. English common figure skates complete with straps.

These skates are not real hard to find, and can be seen at flea markets, antique shows, and malls.

CLASS D $200 - $350

1. Handmade or factory made wooden platform skates with a small sized, (2" dia.) curled front runner and an embellished tip of some type, such as brass or double forged over.
2. Handmade all metal skates with leathers attached.
3. Common child's skates complete with leathers.
4. Early Dutch racing skates with sandwiched blade between wooden platform and with original leathers.

Rarely seen at flea markets but available at antique shows and malls.

CLASS E $350 - $500

1. Handmade or factory made quality wood platforms such as cherry or walnut with medium sized (4" dia.) curled front runner and embellished tip of some kind, such as a brass acorn.
2. Pair of skates with heel enclosed with leather and wide toe leather straps, brass trimmed, with medium sized curl and unusual wooden platform.
3. Early Dutch handmade skates with brass tip and blade sandwiched between wooden curls with original straps.
4. Unusual child's pair of skates complete with original straps and original paint.
5. Solid brass runner and platform skates with engraving on the surfaces, and leather straps riveted on.
6. Any unusual shaped platforms and runners combined with original leathers such as spring blade with no support stanchion.
7, Hand forged medium sized (4" dia.) curl that is an enclosed tight curl or open reverse curl with original leathers and quality wood platform.

These skates are rarely seen at flea markets but a few pop up at the antique shows and malls.

CLASS F $500 - $750

1. Hand or factory made high quality wooden platforms such as bird's eye maple, tiger stripe maple, rosewood, or walnut combined with a nice large curled runner and original leather straps.
2. Exceptionally large curl (5" dia. or larger) with a brass acorn or similar runner tip embellishment mounted on a quality wood platform such as the E.W. Wirth skate made in Germany.
3. All metal forged platform and hand wrought runner combined with copper riveted original leather straps.
4. Large hand forged curled skates wrought from an old worn out file, the tang forming the curl and nicely mounted on a wooden platform with original red paint and leathers.
5. A similar pair as above but forged from an old worn out rasp, the tang forming a large curl attached to a wooden platform with original paint visible.
6. Brass bound leg supported skates with original leathers such as the Blondin Patent skate. Very rarely seen at flea markets, but occasionally a pair will show up at antique shows and markets.

CLASS G $750 - $1000

1. A pair of outstanding blacksmith made, hand forged skates with a large curl (5" diameter) and nice finial at the tip. Superior blacksmith work on all the stanchion and weld joints.

CLASS H $1000 - $1500

1. A Pair of skates with a super big curl forged on the runner as shown in figure 6.
2. A pair of outstanding blacksmith made hand forged skates with a large curl made by twisting the iron for the curl before the curl is forged. The runner blades are highly decorated with engraved diamond shaped figures, and flowers culminating at the curl by a forged diamond crown tip. The wooden platforms are made from walnut and the heel leathers are enclosed with original hand stitched leather. One of a kind super quality handmade pair of skates that a collector generally will not run across. See the cover and figure 88.

CLASS I $1500 - $2000

1. A pair of outstanding hand wrought Swan Skates with long graceful necks as shown in the manuscript. This is a one of a kind handmade pair of ice skates with great lines.

These examples of skates are generally found in museums today or a few private collections.

Many of the American skate designs have European influence in them and are difficult to identify in some cases, because of the early immigration to our country. That is why some of the American skates will have a "European Look ."

Skater's Lanterns

The brief general price guide below includes a dollar range based on a traditional skater's style lantern that is 100% complete with no dents, and only normal wear. Obviously most of the lanterns that have been around for a hundred years have a few small dents in them or other minor problems and would sell for less than the prices listed below. The guide is broken down into two ranges, namely tin and brass with different colored globes. The traditional style tin lantern in a clear glass globe is the most common and least expensive. The colored glass globes are much rarer and command much higher prices. They vary in price depending upon the geographical area of the country as well.

The lanterns are placed in order of their rarity, as I see them. Again everyone has his/her idea of rarity and value. The prices were developed based upon dealer' s tagged prices, confirmed sale prices by dealers, and purchasing lanterns on the market myself.

All other nontraditional skater lantern designs were not included because they vary so much in design and rarity. The skater's lantern prices have increased appreciably in the last few years.

Tin Skater's Lanterns General Price Guide

	Range
1. Tin lantern with clear glass globe	$50 - $150
2. Tin lantern with green globe	$350 - $550
3. Tin lantern with cobalt blue globe	$350 - $550
4. Tin lantern with purple globe	$550 - $750
5. Tin lantern with red globe	$650 - $850
6. Tin lantern with amber globe	$750 - $950

Brass skater's lanterns General Price Guide

	Range
1. Brass lantern with clear globe	$100 - $250
2. Brass lantern with green globe	$500 - $700
3. Brass lantern with cobalt blue globe	$600 - $800
4. Brass lantern with purple globe	$800 - $1000
5. Brass lantern with red globe	$900 - $1200
6. Brass lantern with amber globe	$1000 - $1300

Chapter Twelve

Conclusion

Writing this book has been very challenging, since it is the first of its kind. Identifying the country of origin and establishing a price guide for all the skates were the two most difficult tasks. Because early immigration to our country caused many of the American skate designs to have European influences in them, many of the skates were hard to identify. They have that "European Look."

Ice skate collecting is an unlimited collecting field. Though I have acquired both manufactured and handmade skates, I prefer the handmade, one-of-a-kind skates because of their superior craftsmanship. In addition, handmade skates are not bound by the number of models or styles made, as are the factory made skates. But as a whole, antique ice skate collecting has unlimited possibilities in numbers, as well as variety. Literally thousands of people over centuries have had a hand in making and improving skate designs.

Like most collectors of artifacts, one can never be content with his/her collection. There is always that burning desire to run across another outstanding pair of skates to add to the collection. Antique ice skate collecting is exciting and rewarding and becomes an addiction for some of us, or rather a disease. Generally collectors collect certain items because of the appeal or craftsmanship of an item, rather than for an investment. But I feel antique skate collecting is also a good investment to make, as the prices keep going up.

I trust that this first book on collecting ice skates and skater's lanterns will encourage you to continue your search and enhance your collection. It should provide some guidance and help on identifying, dating, and establishing some price guide ranges for skates. It provides a place to start for collectors and dealers to build the historical picture of antique ice skates. Hopefully by the end, this book will precipitate an Antique Ice Skater's Club for the mutual benefit of the collectors around the world.

Bibliography

Bass, Howard. *Skating Elegance On Ice*. London: Marshall Cavendish Books Limited, 1980.

Brokaw, Irving. *The Art of Skating*. New York: American Sports Publishing Co., 1915.

Brokaw, Irving. *The Art of Skating*. New York: Charles Scribner's Sons, 1926.

Heathcote, J.M., and C.G. Tebbutt. *Skating*. London: Longmans, Green and Co., 1892.

Heller, Mark. *The Illustrated Encyclopedia of Ice Skating*. New York: Paddington Press Ltd., 1979.

Jones, Ernest. *The Elements of Figure Skating*. London: Unwin Brothers Limited, 1931 & 1952.

Dixon, Lauinda. *Skating in the Arts of 17th Century Holland*. Cincinnati, OH: Taft Museum, 1987.

Lambert, Luna. *The American Skating Mania*. National Museum of History and Technology, Smithsonian Institution, Washington, D.C. , December 1978 – February 1979.

Whedon, Julia. *The Fine Art of Ice Skating*. New York: Harry N. Abrams, 1988.